ALEKS

Math Test Preparation

and

study guide

**The Most Comprehensive Prep
Book with Two Full-Length
ALEKS Math Tests**

By

Michael Smith & Reza Nazari

ALEKS Math Test Preparation and Study Guide

Published in the United State of America By

The Math Notion

Email: info@Mathnotion.com

Web: www.MathNotion.com

About the Author

Michael Smith has been a math instructor for over a decade now. He holds a master's degree in Management. Since 2006, Michael has devoted his time to both teaching and developing exceptional math learning materials. As a Math instructor and test prep expert, Michael has worked with thousands of students. He has used the feedback of his students to develop a unique study program that can be used by students to drastically improve their math score fast and effectively.

– GRE Math Test Preparation and Study Guide

– SAT Math Test Preparation and Study Guide

– ACT Math Test Preparation and Study Guide

– Accuplacer Math Workbook

– TSI Math Workbook

– Common Core Math Workbooks

– GED Math Workbook

– and many Math Education Workbooks…

As an experienced Math teacher, Mr. Smith employs a variety of formats to help students achieve their goals: He tutors online and in person, he teaches students in large groups, and he provides training materials and textbooks through his website and through Amazon.

You can contact Michael via email at:

info@Mathnotion.com

ALEKS Math Test Preparation and Study Guide

ALEKS Math Test Preparation and Study Guide covers all mathematics topics that will be key to succeeding on the ALEKS Math test. The step-by-step guide and hundreds of examples in this book can help you hone your math skills, boost your confidence, and be well prepared for the ALEKS test on test day.

Inside the pages of this comprehensive ALEKS Math prep book, you can learn math topics in a structured manner with a complete study program to help you understand essential math skills. Two full-length ALEKS Math practice tests and more than 2,000 ALEKS math questions will help you find your weak areas, raise your score, and beat the ALEKS Math Test.

ALEKS Math Test Preparation and Study Guide contains many exciting features to help you ace the ALEKS Math test, including:

- Content 100% aligned with the 2019 ALEKS® test
- 2,000+ ALEKS Math practice questions with answers
- Provided and tested by ALEKS Math test experts
- Complete coverage of all ALEKS Math topics which you will be tested
- Many Math skill building exercises to help you approach different Math question types
- Exercises on different ALEKS Math topics such as numbers, ratios, percent, equations, polynomials, exponents and geometry
- 2 full-length practice tests (featuring new question types) with detailed answers

The only prep book you will ever need to ace the ALEKS Math Test!

WWW.MathNotion.COM

… So Much More Online!

✓ FREE Math Lessons

✓ More Math Learning Books!

✓ Mathematics Worksheets

✓ Online Math Tutors

For a PDF Version of This Book

Please Visit www.MathNotion.com

Contents

Chapter 1:

Integer and Complex Numbers

Topics that you'll learn in this chapter:

- ➢ Rounding and Estimates
- ➢ Addition and Subtraction Integers
- ➢ Multiplication and Division Integers
- ➢ Arrange and ordering Integers and Numbers
- ➢ Comparing Integers, Order of Operations
- ➢ Mixed Integer Computations
- ➢ Integers and Absolute Value
- ➢ Adding and Subtracting Complex Numbers
- ➢ Multiplying and Dividing Complex Numbers
- ➢ Graphing Complex Numbers
- ➢ Rationalizing Imaginary Denominators

"Wherever there is number, there is beauty." –Proclus

Unlock problems

❖ Rounding

✓ Rounding is putting a number up or down to the nearest whole number by rounding 1 to 4 down and 5 to 9 up.

Example: 64 rounded to the nearest ten is 60, because 4 is less than 5 and round down to 60 rather than 70.

❖ Rounding and Estimates

✓ Rounding and estimating are math strategies used for approximating a number.

✓ To estimate means to make a rough guess or calculation.

✓ To round means to simplify a known number by scaling it slightly up or down.

Example: $73 + 69 \approx 70 + 70 = 140$

❖ Whole Number Addition and Subtraction

✓ Line up the numbers.

✓ Start with the unit place (ones place).

✓ Regroup if necessary.

✓ Add or subtract the tens place.

✓ Continue with other digits.

Example:

$$231 + 120 = 35$$

$$292 - 90 = 202$$

❖ Whole Number Multiplication and Division

Multiplication:

- ✓ Learn the times tables first!
- ✓ For multiplication, line up the numbers you are multiplying.
- ✓ Start with the ones place.
- ✓ Continue with other digits

Division:

- ✓ A typical division problem: Dividend ÷ Divisor = Quotient
- ✓ In division, we want to find how many times a number (divisor) is contained in another number (dividend).
- ✓ The result in a division problem is the quotient.

Example:

$$200 \times 90 = 18,000$$
$$18,000 \div 90 = 200$$

❖ Adding and Subtracting Integers

- ✓ Integers: $\{\dots, -3, -2, -1, 0, 1, 2, 3, \dots\}$
- ✓ Includes: zero, counting numbers, and the negative of the counting numbers.
- ✓ Add a positive integer by moving to the right on the number line.
- ✓ Add a negative integer by moving to the left on the number line.
- ✓ Subtract an integer by adding its opposite.

Example:

$$7 + 16 = 23$$

$$15 + (-13) = 2$$

$$(-34) + 14 = -20$$

$$(-14) + (-12) = -26$$

$$14 - (-13) = 27$$

❖ Multiplying and Dividing Integers

✓ (positive) × (positive) = positive

✓ (positive) ÷ (positive) = positive

✓ (negative) × (negative) = positive

✓ (negative) ÷ (negative) = positive

✓ (negative) × (positive) = negative

✓ (negative) ÷ (positive) = negative

✓ (positive) × (negative) = negative

✓ (positive) ÷ (negative) = negative

×⁄÷	+	−
+	+	−
−	−	+

Examples:

$$(+7) \times (+2) = 7 \times 2 = 14$$

$$(+3) \times (-5) = (-5) \times (+3) = -15$$

$$(-2) \times (-2) = (+4) = 4$$

$$(+10) \div (+2) = 10 \div 2 = 5$$

$$(-4) \div (+2) = -2$$

$$(-12) \div (-6) = 3$$

❖ Arrange, Order, and Comparing Integers

✓ When using a number line, numbers increase as you move to

the right.

Examples:

$5 < 7,$

$-5 < -2$

$-18 < -12$

✓ To compare numbers, you can use number line! As you move from left to right on the number line, you find a bigger number!

Example:

Order integers from least to greatest.

$(-11, -13, 7, -2, 12)$

$-13 < -11 < -2 < 7 < 12$

❖ Order of Operations

✓ Use "order of operations" rule when there are more than one math operation.

✓ PEMDAS (parentheses/ exponents/ multiply/ divide/ add/ subtract)

Example:

$(12 + 4) \div (-4) = -4$

❖ Integers and Absolute Value

✓ To find an absolute value of a number, just find its distance from 0 (Zero).

Example:

$|-6| = 6$

$|6| = 6$

$|-12| = 12$

$|12| = 12$

❖ Adding and Subtracting Complex Numbers

✓ Adding:

$$(a + bi) + (c + di) = (a + c) + (b + d)i$$

✓ Subtracting:

$$(a + bi) - (c + di) = (a - c) + (b - d)i$$

Example:

$$-5 + (2 - 4i) = -3 - 4i$$

$$(2 - 5i) + (4 - 6i) = 6 - 11i$$

❖ Multiplying and Dividing Complex Numbers

✓ Multiplying:

$$(a + bi) + (c + di) = (ac - bd) + (ad + bc)i$$

✓ Dividing:

$$\frac{a+bi}{c+di} = \frac{a+bi}{c+di} \cdot \frac{c-di}{c-di} = \frac{ac+bd}{c^2 + d^2} + \frac{bc+ad}{c^2 + d^2}i$$

❖ Graphing Complex Numbers

✓ Complex numbers can be plotted on the complex coordinate plane.

✓ The horizontal line is Real axis and the vertical line is Imaginary axis.

✓ Complex numbers are written in the form of: $A + Bi$, where

A is real number and B is number of units up or down.

Example: The point 3 + 4i, is located 3 units to the right of origin and 4 units up.

❖ Rationalizing Imaginary Denominators

✓ Step 1: Find the conjugate (it's the denominator with different sign between the two terms.

✓ Step 2: Multiply numerator and denominator by the conjugate.

✓ Step 3: Simplify if needed.

Example:

$$\frac{5i}{2-3i} = \frac{5i(2+3i)}{(2-3i)(2+3i)} = \frac{10i+15i^2}{4-9i^2} = \frac{-15+10i}{13}$$

Adding and Subtracting Integers

✎ **Find the sum.**

1) $(-14) + (-5)$

2) $7 + (-21)$

3) $(-15) + 24$

4) $(-9) + 28$

5) $33 + (-14)$

6) $(-23) + (-4) + 3$

7) $3 + (-16) + (-20) + (-19)$

8) $(-28) + (-19) + 31 + 16$

9) $(-7) + (-11) + (27 - 19)$

10) $6 + (-20) + (35 - 24)$

11) $(+24) + (+32) + (-47)$

12) $41 + 17 + (-29)$

✎ **Find the difference.**

13) $(-6) - (-32)$

14) $(-14) - (9)$

15) $(26) - (-8)$

16) $(42) - (7)$

17) $(-13) - (-7) - (19)$

18) $(64) - (-3) + (-6)$

19) $(7) - (4) - (-2)$

20) $(3) - (5) - (-14)$

21) $(24) - (3) - (-24)$

22) $(-37) - (-72)$

23) $(-11) - 24 + 32$

24) $32 - (-16) - (-13)$

Multiplying and Dividing Integers

✎ Find each product.

1) $(-7) \times (-4)$

2) 7×6

3) $(-3) \times 7 \times (-4)$

4) $3 \times (-7) \times (-7)$

5) $12 \times (-14)$

6) $20 \times (-6)$

7) 9×8

8) $(-5) \times (-13)$

9) $7 \times (-8) \times 3$

10) $8 \times (-1) \times 5$

11) $(-7) \times (-9)$

12) $(-12) \times (-11) \times 2$

✎ Find each quotient.

13) $56 \div 8$

14) $(-60) \div 4$

15) $(-72) \div (-8)$

16) $28 \div (-7)$

17) $38 \div (-2)$

18) $(-84) \div (-12)$

19) $37 \div (-1)$

20) $(-169) \div 13$

21) $81 \div 9$

22) $(-24) \div (-3)$

23) $(-6) \div (-1)$

24) $(-65) \div 5$

Arrange, Order, and Comparing Integers

✍ **Order each set of integers from least to greatest.**

1) $-12, -17, 12, -1, 1$ ____, ____, ____, ____, ____, ____

2) $11, -7, 5, -3, 2$ ____, ____, ____, ____, ____, ____

3) $25, -52, 19, 0, -22$ ____, ____, ____, ____, ____, ____

4) $31, -84, 0, -13, 47, -55$ ____, ____, ____, ____, ____, ____

5) $-45, 39, 21, -18, -51, 42$ ____, ____, ____, ____, ____, ____

6) $-17, -65, 71, -25, -51, -39$ ____, ____, ____, ____, ____, ____

✍ **Order each set of integers from greatest to least.**

7) $81, 5, 36, 19, 77, 24$ ____, ____, ____, ____, ____, ____

8) $-1, 7, -3, 4, -7$ ____, ____, ____, ____, ____, ____

9) $-47, 17, -17, 27, 37$ ____, ____, ____, ____, ____, ____

10) $-21, 19, -14, -17, 15$ ____, ____, ____, ____, ____, ____

11) $1, 0, -1, -2, 2, -3$ ____, ____, ____, ____, ____, ____

12) $-124, -91, 31, -28, -75, 19$ ____, ____, ____, ____, ____, ____

✎ Compare. Use >, =, <

1) $0 \underline{\quad} 1$

2) $-12 \underline{\quad} -17$

3) $0 \underline{\quad} -21$

4) $41 \underline{\quad} -56$

5) $-654 \underline{\quad} -645$

6) $-42 \underline{\quad} -48$

7) $-68 \underline{\quad} -20$

8) $-86 \underline{\quad} -106$

9) $-26 \underline{\quad} (-26)$

10) $425 \underline{\quad} -425$

Order of Operations

✎ Evaluate each expression.

1) $41 - (8 \times 3)$

2) $7 \times 6 - (\dfrac{16}{12 - (-4)})$

3) $32 - (4 \times (-2))$

4) $(6 \times 5) + (-3)$

5) $(\dfrac{(-2)+4}{(-1)+(-1)}) \times (-6)$

6) $(14 + (-2) - 3) \times 7 - 5$

7) $\dfrac{40}{3\,(9 - (-1)) - 10}$

8) $38 - (4 \times 6)$

9) $-43 + (4 \times 8)$

10) $((-12) + 18) \div (-2)$

11) $(-60 \div 3) \div (-12 - 8)$

12) $47 + (-8) \times (\dfrac{(-18)}{6})$

Integers and Absolute Value

✎ **Write absolute value of each number.**

1) 22

2) − 12

3) − 31

4) 0

5) 47

6) − 9

7) − 1

8) 37

9) -23

10) − 4

11) − 57

12) 19

13) − 15

14) − 55

✎ **Evaluate.**

15) $|-29| - |13| + 20$

16) $39 + |-15 - 42| - |3|$

17) $28 - |-47| - 61$

18) $|56| - |-18| + 19$

19) $|101| - |-38| - 20$

20) $|42| - |-68| + 70$

21) $|-87 + 73| + 15 - 9$

22) $|-6| + |-17|$

23) $|-9 + 5 - 2| + |6 + 6|$

24) $|-14| - |-23| - 5$

Adding and Subtracting Complex Numbers

✎ *Simplify.*

1) $-6 + (3i) + (-6 + 5i)$

2) $10 - (6i) + (5 - 12i)$

3) $-3 + (-5 - 6i) - 8$

4) $(-15 - 4i) + (12 + 6i)$

5) $(4 + 2i) + (9 + 3i)$

6) $(6 - 2i) + (3 + i)$

7) $3 + (4 - 4i)$

8) $(9 + 9i) + (6 + 5i)$

9) $(-4i) - (-6 + 2i)$

10) $(-12 + 2i) - (-10 - 10i)$

11) $(-12i) + (2 - 4i) + 8$

12) $(-10 - 8i) - (-8 - 2i)$

13) $(13i) - (15 + 3i)$

14) $(-2 + 4i) - (-6 - i)$

15) $(-3 + 15i) - (-5 + 5i)$

16) $(-12i) + (3 - 4i) + 5$

Multiplying and Dividing Complex Numbers

✎ *Simplify.*

1) $(2i)(-i)(2-5i)$

2) $(2-5i)(2-4i)$

3) $(-3+6i)(2+5i)$

4) $(5+3i)(5+8i)$

5) $(2+3i)^2$

6) $3(3i)-(2i)(-5+3i)$

7) $\dfrac{3+2i}{12+2i}$

8) $\dfrac{2-2i}{-3i}$

9) $\dfrac{2+6i}{-1+8i}$

10) $\dfrac{-5+i}{-7+i}$

11) $\dfrac{4+5i}{i}$

12) $\dfrac{-2i}{4-2i}$

13) $\dfrac{2}{-9i}$

14) $\dfrac{-2-6i}{4i}$

15) $\dfrac{9i}{3-i}$

16) $\dfrac{-1+3i}{-6-5i}$

17) $\dfrac{-2-4i}{-2+3i}$

18) $\dfrac{6+i}{2-7i}$

Graphing Complex Numbers

Identify each complex number graphed.

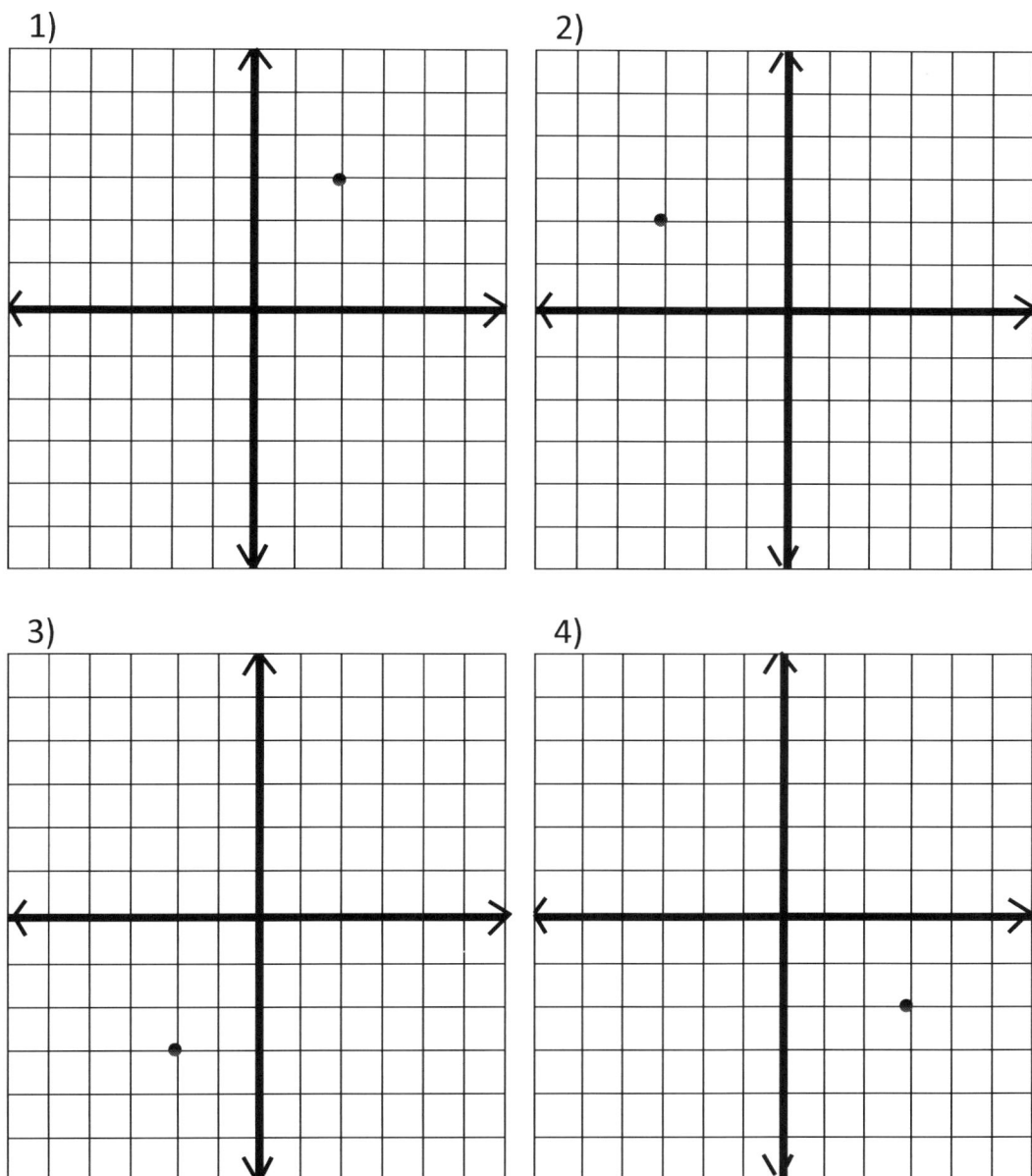

1)

2)

3)

4)

Rationalizing Imaginary Denominators

✎ *Simplify.*

1) $\dfrac{8 - 8i}{-4i}$

2) $\dfrac{2 - 10i}{-8i}$

3) $\dfrac{5 + 7i}{9i}$

4) $\dfrac{6i}{-1+2i}$

5) $\dfrac{7i}{-2 - 6i}$

6) $\dfrac{-10 - 5i}{-4 + 4i}$

7) $\dfrac{-2 - 6i}{6 + 8i}$

8) $\dfrac{-6-3i}{7-12i}$

9) $\dfrac{-1 + i}{-7i}$

10) $\dfrac{-4 - i}{i}$

11) $\dfrac{c}{ib}$

12) $\dfrac{-3 - i}{6 + 4i}$

13) $\dfrac{-5 + i}{-3i}$

14) $\dfrac{-8}{-i}$

15) $\dfrac{-4 - i}{-1 + 4i}$

16) $\dfrac{-16- 4i}{-4 + 4i}$

17) $\dfrac{8}{-6i}$

18) $\dfrac{-4i-1}{-1 + 3i}$

Answers of Worksheets – Chapter 1

Adding and Subtracting Integers

1) − 19	9) − 10	17) − 25
2) − 14	10) − 3	18) 61
3) 9	11) 9	19) 5
4) 19	12) 29	20) 12
5) 19	13) 26	21) 45
6) − 24	14) −23	22) 35
7) − 52	15) 34	23) −3
8) 0	16) 35	24) 61

Multiplying and Dividing Integers

1) 28	9) − 168	17) −19
2) 42	10) − 40	18) 7
3) 84	11) 63	19) −37
4) 147	12) 264	20) −13
5) − 168	13) 7	21) 9
6) − 120	14) − 15	22) 8
7) 72	15) 9	23) 6
8) 65	16) − 4	24) −13

Arrange and Order, Comparing Integers

1) − 17, − 12, − 1, 1, 12	7) 81, 77, 36, 24, 19, 5
2) − 7, − 3, 2, 5, 11	8) 7, 4, − 1, − 3, − 7
3) − 52, − 22, 0, 19, 25	9) 37, 27, 17, − 17, − 47
4) − 84, − 55, − 13, 0, 31, 47	10) 19, 15, − 14, − 17, −21
5) −51, −45, −18, 21, 39, 42	11) −3, −2, -1, 0,1,2
6) − 65, − 51, − 39, − 25, − 17, 71	12) 31, 19, −28, −75, −91, −124

Compare.

1) <	5) <	9) =
2) >	6) >	10) >
3) >	7) <	
4) >	8) >	

Order of Operations

1) 17	5) 6	9) −11
2) 41	6) 58	10) −3
3) 40	7) 2	11) 1
4) 27	8) 14	12) 71

Integers and Absolute Value

1) 22	9) 23	17) − 80
2) 12	10) 4	18) 57
3) 31	11) 57	19) 43
4) 0	12) 19	20) 44
5) 47	13) 15	21) 20
6) 9	14) 55	22) 23
7) 1	15) 36	23) 18
8) 37	16) 93	24) −14

Adding and Subtracting Complex Numbers

1) − 12 + 8i	7) 7 − 4i	13) − 15 + 10i
2) 15− 18i	8) 15 + 14i	14) 4 + 5i
3) −16 − 6i	9) 6 − 6i	15) 2 + 10i
4) −3 + 2i	10) −2 + 12i	16) 8 − 16i
5) 13 + 5i	11) 10 − 16i	
6) 9 − i	12) −2 − 6i	

Multiplying and Dividing Complex Numbers

1) $4 - 10i$

2) $-16 - 18i$

3) $-36 - 3i$

4) $1 + 55i$

5) $-5 + 12i$

6) $6 + 19i$

7) $\frac{40 + 18i}{148}$

8) $\frac{2}{3} + \frac{2}{3}i$

9) $\frac{46}{65} - \frac{22}{65}i$

10) $\frac{18 - i}{25}$

11) $-4i + 5$

12) $\frac{1}{5} - \frac{2}{5}i$

13) $\frac{2i}{9}$

14) $\frac{i - 3}{2}$

15) $\frac{27i - 9}{10}$

16) $-\frac{9}{61} - \frac{23i}{61}$

17) $-\frac{8}{13} + \frac{14i}{13}$

18) $\frac{5}{53} + \frac{44i}{53}$

Graphing Complex Numbers

1) $2 + 3i$

2) $-3 + 2i$

3) $-2 - 3i$

4) $3 - 2i$

Rationalizing Imaginary Denominators

1) $2i + 2$

2) $\frac{5 + i}{4}$

3) $\frac{-7 + 5i}{9}$

4) $\frac{-6i + 12}{5}$

5) $\frac{-7i - 21}{20}$

6) $\frac{5 + 15i}{8}$

7) $\frac{-3 - i}{5}$

8) $\frac{-6 - 93i}{193}$

9) $\frac{-i - 1}{7}$

10) $4i - 1$

11) $-\frac{ic}{b}$

12) $\frac{-11 + 3i}{26}$

13) $\frac{-1 - 5i}{3}$

14) $-8i$

15) $0 + 1i$

16) $\frac{3 + 5i}{2}$

17) $\frac{4i}{3}$

18) $\frac{-11 + 7i}{10}$

Chapter 2: Fractions and Decimals

Topics that you'll learn in this chapter:

➢ Simplifying Fractions

➢ Adding and Subtracting Fractions, Mixed Numbers and Decimals

➢ Multiplying and Dividing Fractions, Mixed Numbers and Decimals

➢ Comparing and Rounding Decimals

➢ Converting Between Fractions, Decimals and Mixed Numbers

➢ Factoring Numbers, Greatest Common Factor, and Least Common Multiple

➢ Divisibility Rules

"A Man is like a fraction whose numerator is what he is and whose denominator is what he thinks of himself. The larger the denominator, the smaller the fraction." –Tolstoy

Unlock problems

❖ <u>Simplifying Fractions:</u>

✓ Evenly divide both the top and bottom of the fraction by $2, 3, 5, 7, \ldots$ etc.

✓ Continue until you can't go any further.

Example: $\dfrac{4}{12} = \dfrac{2}{6} = \dfrac{1}{3}$

❖ <u>Factoring Numbers</u>

✓ Factoring numbers means to break the numbers into their prime factors.

✓ First few prime numbers: $2, 3, 5, 7, 11, 13, 17, 19$

Example: $12 = 2 \times 2 \times 3$

❖ <u>Greatest Common Factor (GCF)</u>

✓ List the prime factors of each number.

✓ Multiply common prime factors.

Example:

$200 = 2 \times 2 \times 2 \times 5 \times 5$

$60 = 2 \times 2 \times 3 \times 5$

$GCF\,(200, 60) = 2 \times 2 \times 5 = 20$

❖ <u>Least Common Multiple (LCM)</u>

✓ Find the GCF for the two numbers.

✓ Divide that GCF into either number.

✓ Take that answer and multiply it by the other number.

Example:

LCM $(200, 60)$:

GCF is 20

$200 \div 20 = 10$

$10 \times 60 = 600$

❖ <u>Divisibility Rules</u>

✓ Divisibility means that a number can be divided by other numbers evenly.

Example: 24 is divisible by 6, because $24 \div 6 = 4$

❖ <u>Adding and Subtracting Fractions</u>

✓ For "like" fractions (fractions with the same denominator), add or subtract the numerators and write the answer over the common denominator.

✓ Find equivalent fractions with the same denominator before you can add or subtract fractions with different denominators.

✓ Adding and Subtracting with the same denominator:

$$\frac{a}{b} + \frac{c}{b} = \frac{a+c}{b}$$

$$\frac{a}{b} - \frac{c}{b} = \frac{a-c}{b}$$

✓ Adding and Subtracting fractions with different denominators:

$$\frac{a}{b} + \frac{c}{d} = \frac{ad+cb}{bd}$$

$$\frac{a}{b} - \frac{c}{d} = \frac{ad - cb}{bd}$$

❖ Multiplying and Dividing Fractions

✓ **Multiplying fractions**: multiply the top numbers and multiply the bottom numbers.

Example: $\frac{a}{b} \times \frac{c}{d} = \frac{a \times c}{b \times d}$

✓ **Dividing fractions**: Keep, Change, Flip Keep first fraction, change division sign to multiplication, and flip the numerator and denominator of the second fraction. Then, solve!

Example: $\frac{a}{b} \div \frac{c}{d} = \frac{a}{b} \times \frac{d}{c} = \frac{ad}{bc}$

❖ Adding Mixed Numbers

Use the following steps for both adding and subtracting mixed numbers.

✓ Find the Least Common Denominator (LCD)

✓ Find the equivalent fractions for each mixed number.

✓ Add fractions after finding common denominator.

✓ Write your answer in lowest terms.

Example: $1\frac{3}{4} + 2\frac{3}{8} = 4\frac{1}{8}$

❖ Subtract Mixed Numbers

Use the following steps for both adding and subtracting mixed numbers.

✓ Find the Least Common Denominator (LCD)

✓ Find the equivalent fractions for each mixed number.

✓ Add or subtract fractions after finding common denominator.

✓ Write your answer in lowest terms.

Example: $5\frac{2}{3} - 3\frac{2}{7} = 2\frac{8}{21}$

❖ Multiplying Mixed Numbers

✓ Convert the mixed numbers to improper fractions.

✓ Multiply fractions and simplify if necessary.

$$a\frac{c}{b} = a + \frac{c}{b} = \frac{ab+c}{b}$$

Example:

$$2\frac{1}{3} \times 5\frac{3}{7} = \frac{7}{3} \times \frac{38}{7} = \frac{38}{3} = 12\frac{2}{3}$$

❖ Dividing Mixed Numbers

✓ Convert the mixed numbers to improper fractions.

✓ Divide fractions and simplify if necessary.

$$a\frac{c}{b} = a + \frac{c}{b} = \frac{ab+c}{b}$$

Example:

$$10\frac{1}{2} \div 5\frac{3}{5} = \frac{21}{2} \div \frac{28}{5} = \frac{21}{2} \times \frac{5}{28} = \frac{105}{56} = 1\frac{7}{8}$$

❖ Comparing Decimals

✓ **Decimals:** is a fraction written in a special form. For example, instead of writing $\frac{1}{2}$ you can write 0.5.

For comparing:

Equal to $=$; Less than $<$; greater than $>$

greater than or equal $\geq$; Less than or equal $\leq$

Example:

$2.67 > 0.267$

❖ Rounding Decimals

✓ We can round decimals to a certain accuracy or number of decimal places. This is used to make calculation easier to do and results easier to understand, when exact values are not too important.

First, you'll need to remember your place values:

12.4567

1: tens	2: ones	4: tenths
5: hundredths	6: thousandths	7: ten thousandths

Example:

$\underline{6}.37 = 6$

❖ Adding and Subtracting Decimals

✓ Line up the numbers.

✓ Add zeros to have same number of digits for both numbers.

✓ Add or Subtract using column addition or subtraction.

Example:

$$\begin{array}{r} 16.18 \\ -\ 13.45 \\ \hline 2.73 \end{array}$$

❖ <u>Multiplying and Dividing Decimals</u>

For Multiplication:

- ✓ Set up and multiply the numbers as you do with whole numbers.
- ✓ Count the total number of decimal places in both of the factors.
- ✓ Place the decimal point in the product.

For Division:

- ✓ If the divisor is not a whole number, move decimal point to right to make it a whole number. Do the same for dividend.
- ✓ Divide similar to whole numbers.

❖ <u>Converting Between Fractions, Decimals and Mixed Numbers</u>

Fraction to Decimal:

- ✓ Divide the top number by the bottom number.

Decimal to Fraction:

- ✓ Write decimal over 1.
- ✓ Multiply both top and bottom by 10 for every digit on the right side of the decimal point.
- ✓ Simplify.

Simplifying Fractions

✎ **Simplify the fractions.**

1) $\dfrac{42}{62}$

2) $\dfrac{18}{24}$

3) $\dfrac{10}{15}$

4) $\dfrac{36}{48}$

5) $\dfrac{9}{27}$

6) $\dfrac{20}{80}$

7) $\dfrac{12}{27}$

8) $\dfrac{28}{56}$

9) $\dfrac{40}{100}$

10) $\dfrac{7}{63}$

11) $\dfrac{25}{45}$

12) $\dfrac{24}{32}$

13) $5\dfrac{35}{56}$

14) $2\dfrac{26}{32}$

15) $9\dfrac{5}{25}$

16) $3\dfrac{35}{56}$

17) $1\dfrac{52}{104}$

18) $4\dfrac{13}{65}$

19) $1\dfrac{45}{60}$

20) $\dfrac{54}{60}$

21) $7\dfrac{66}{132}$

Factoring Numbers

✍ **List all positive factors of each number.**

1) 32

2) 26

3) 40

4) 96

5) 60

6) 28

7) 35

8) 64

9) 56

10) 75

11) 81

12) 48

✍ **List the prime factorization for each number.**

13) 20

14) 65

15) 99

16) 42

17) 84

18) 30

19) 52

20) 105

21) 75

22) 48

23) 36

24) 31

25) 24

26) 54

Greatest Common Factor (GCF)

⊠ Find the GCF for each number pair.

1) 14, 28

2) 48, 36

3) 6, 18

4) 25, 15

5) 52, 39

6) 57, 38

7) 16, 12

8) 45, 60

9) 36, 72

10) 27, 63

11) 64, 48

12) 80, 45

13) 36, 42

14) 15, 90

15) 63, 84

16) 100, 75

17) 26, 42

18) 93, 62

Least Common Multiple (LCM)

⊠ Find the LCM for each number pair.

1) 6, 9

2) 25, 35

3) 64, 48

4) 12, 18

5) 14, 21

6) 45, 15

7) 42, 63

8) 21, 12

9) 64, 44

10) 15, 12

11) 75, 6

12) 20, 10, 40

13) 16, 32, 24

14) 15, 25, 35

15) 14, 8, 21

16) 5, 9, 7

17) 14, 6, 16

18) 36, 60, 24

Divisibility Rules

✏️ Use the divisibility rules to find the factors of each number

1) 24   2 3 4 5 6 7 8 9 10

2) 32   2 3 4 5 6 7 8 9 10

3) 16   2 3 4 5 6 7 8 9 10

4) 42   2 3 4 5 6 7 8 9 10

5) 28   2 3 4 5 6 7 8 9 10

6) 56   2 3 4 5 6 7 8 9 10

7) 48   2 3 4 5 6 7 8 9 10

8) 36   2 3 4 5 6 7 8 9 10

9) 81   2 3 4 5 6 7 8 9 10

10) 50   2 3 4 5 6 7 8 9 10

11) 63   2 3 4 5 6 7 8 9 10

12) 84   2 3 4 5 6 7 8 9 10

Adding and Subtracting Fractions

✎ **Add fractions.**

1) $\dfrac{1}{3} + \dfrac{1}{2}$

4) $\dfrac{5}{12} + \dfrac{1}{3}$

7) $\dfrac{2}{7} + \dfrac{5}{7}$

2) $\dfrac{2}{7} + \dfrac{2}{3}$

5) $\dfrac{3}{6} + \dfrac{1}{5}$

8) $\dfrac{5}{13} + \dfrac{2}{4}$

3) $\dfrac{3}{7} + \dfrac{4}{9}$

6) $\dfrac{3}{15} + \dfrac{2}{5}$

9) $\dfrac{16}{56} + \dfrac{3}{16}$

✎ **Subtract fractions.**

10) $\dfrac{3}{5} - \dfrac{1}{10}$

13) $\dfrac{1}{8} - \dfrac{1}{9}$

16) $\dfrac{2}{25} - \dfrac{1}{15}$

11) $\dfrac{5}{8} - \dfrac{2}{5}$

14) $\dfrac{3}{5} - \dfrac{5}{12}$

17) $\dfrac{3}{4} - \dfrac{13}{18}$

12) $\dfrac{5}{6} - \dfrac{2}{7}$

15) $\dfrac{5}{8} - \dfrac{5}{16}$

18) $\dfrac{8}{42} - \dfrac{7}{48}$

Multiplying and Dividing Fractions

Multiplying fractions. Then simplify.

1) $\frac{2}{7} \times \frac{3}{8}$

2) $\frac{4}{25} \times \frac{5}{8}$

3) $\frac{9}{40} \times \frac{10}{27}$

4) $\frac{6}{13} \times \frac{26}{33}$

5) $\frac{9}{12} \times \frac{4}{5}$

6) $\frac{12}{17} \times \frac{3}{5}$

7) $\frac{28}{115} \times 0$

8) $\frac{8}{9} \times \frac{9}{8}$

9) $\frac{14}{45} \times \frac{15}{28}$

Dividing fractions.

10) $0 \div \frac{1}{10}$

11) $\frac{5}{12} \div 5$

12) $\frac{6}{11} \div \frac{3}{4}$

13) $\frac{21}{32} \div \frac{7}{8}$

14) $\frac{4}{19} \div \frac{8}{19}$

15) $\frac{3}{16} \div \frac{15}{32}$

16) $\frac{5}{7} \div \frac{1}{6}$

17) $\frac{16}{25} \div \frac{4}{5}$

18) $\frac{2}{17} \div \frac{2}{13}$

19) $9 \div \frac{1}{6}$

20) $\frac{12}{28} \div \frac{3}{7}$

21) $\frac{6}{17} \div \frac{5}{14}$

Adding and Subtracting Mixed Numbers

✎ **Add.**

1) $2\frac{1}{4} + 3\frac{1}{4}$

2) $5\frac{3}{4} + 2\frac{1}{4}$

3) $1\frac{1}{9} + 2\frac{2}{9}$

4) $3\frac{1}{6} + 2\frac{2}{3}$

5) $5\frac{4}{15} + 5\frac{3}{10}$

6) $4\frac{1}{7} + 1\frac{1}{3}$

7) $1\frac{5}{21} + 1\frac{5}{28}$

8) $3\frac{2}{5} + 1\frac{3}{7}$

9) $3\frac{3}{8} + 4\frac{1}{12}$

10) $12 + \frac{1}{8}$

11) $2\frac{5}{18} + \frac{5}{24}$

12) $4\frac{7}{16} + 1\frac{1}{2}$

✎ **Subtract.**

1) $4\frac{3}{8} - 1\frac{1}{8}$

2) $3\frac{7}{12} - \frac{1}{3}$

3) $5\frac{9}{14} - 5\frac{6}{21}$

4) $11\frac{4}{9} - 7\frac{1}{4}$

5) $3\frac{1}{3} - 2\frac{2}{3}$

6) $7\frac{1}{8} - 2\frac{1}{2}$

7) $5\frac{4}{17} - 2\frac{4}{17}$

8) $4\frac{10}{11} - 2\frac{1}{3}$

9) $4\frac{2}{7} - 3\frac{2}{5}$

10) $4\frac{5}{9} - 2\frac{7}{18}$

11) $5\frac{7}{15} - 3\frac{2}{15}$

12) $6\frac{1}{21} - 2\frac{1}{35}$

Multiplying and Dividing Mixed Numbers

✎ **Find each product.**

1) $1\frac{1}{4} \times 1\frac{1}{4}$

2) $2\frac{3}{7} \times 1\frac{2}{3}$

3) $4\frac{3}{5} \times 3\frac{3}{4}$

4) $4\frac{1}{8} \times 1\frac{2}{5}$

5) $3\frac{2}{3} \times 2\frac{1}{5}$

6) $1\frac{1}{9} \times 1\frac{2}{3}$

7) $4\frac{1}{2} \times 1\frac{3}{8}$

8) $4\frac{1}{2} \times 2\frac{1}{5}$

9) $2\frac{2}{3} \times 3\frac{1}{4}$

10) $3\frac{1}{5} \times 3\frac{3}{4}$

11) $1\frac{1}{6} \times 1\frac{1}{3}$

12) $1\frac{5}{9} \times 1\frac{4}{5}$

✎ **Find each quotient.**

1) $2\frac{3}{5} \div 1\frac{3}{5}$

2) $4\frac{1}{6} \div 1\frac{2}{3}$

3) $3\frac{2}{5} \div 1\frac{2}{15}$

4) $3\frac{3}{4} \div 2\frac{5}{8}$

5) $1\frac{3}{7} \div 1\frac{1}{2}$

6) $2\frac{5}{6} \div 2\frac{2}{3}$

7) $2\frac{2}{3} \div 3\frac{1}{5}$

8) $3\frac{1}{5} \div 2\frac{2}{3}$

9) $2\frac{1}{7} \div 1\frac{3}{5}$

10) $2\frac{7}{12} \div 6\frac{1}{5}$

11) $3\frac{2}{7} \div 1\frac{5}{9}$

12) $0 \div 1\frac{1}{2}$

Comparing Decimals

✎ **Write the correct comparison symbol (>, < or =).**

1) 1.15 _____ 2.15

2) 0.4 _____ 0.385

3) 12.5 _____ 12.500

4) 4.05 _____ 4.50

5) 0.511 _____ 0.51

6) 0.623 _____ 0.723

7) 8.76 _____ 8.678

8) 3.0069 _____ 3.069

9) 22.042 _____ 22.034

10) 5.11 _____ 5.08

11) 1.11 _____ 1.111

12) 0.06 _____ 0.55

13) 3.204 _____ 3.25

14) 3.92 _____ 3.0952

15) 0.44 _____ 0.044

16) 17.04 _____ 17.040

17) 0.090 _____ 0.80

18) 0.021_____0.201

19) 0.0067_____0.00089

20) 0.79 _____ 0.6

Rounding Decimals

✏ **Round each decimal number to the nearest place indicated.**

1) 0.6̲3

2) 3.0̲4

3) 7.6̲23

4) 0.3̲66

5) 7̲.707

6) 0.08̲8

7) 6.2̲4

8) 76.76̲0

9) 3.62̲9

10) 12.3̲858

11) 1.0̲9

12) 4.2̲57

13) 3.2̲43

14) 6.05̲40

15) 93̲.69

16) 37̲.45

17) 41̲7.078

18) 312.6̲55

19) 18.40̲09

20) 85̲.85

21) 3.20̲8

22) 57.0̲73

23) 126.5̲18

24) 7.00̲69

25) 0.011̲1

26) 11.340̲6

27) 6.3̲891

Adding and Subtracting Decimals

✎ **Add and subtract decimals.**

$$
\begin{array}{r}
37.69 \\
-\ 15.58 \\
\hline
\end{array}
$$
1)

$$
\begin{array}{r}
84.10 \\
-\ 43.45 \\
\hline
\end{array}
$$
4)

$$
\begin{array}{r}
56.93 \\
+\ 23.07 \\
\hline
\end{array}
$$
2)

$$
\begin{array}{r}
121.26 \\
+\ 78.97 \\
\hline
\end{array}
$$
5)

$$
\begin{array}{r}
18.96 \\
+\ 12.87 \\
\hline
\end{array}
$$
3)

$$
\begin{array}{r}
65.00 \\
-\ 53.39 \\
\hline
\end{array}
$$
6)

✎ **Solve.**

7) ___ $+ 3.3 = 5.08$

10) $6.9 -$ ___ $= 0.047$

8) $5.05 +$ ___ $= 14.6$

11) ___ $+ 0.074 = 1.084$

9) $12.9 -$ ___ $= 7.25$

12) ___ $- 6.62 = 31.72$

Multiplying and Dividing Decimals

✎ **Find each product**

1) $\begin{array}{r} 5.5 \\ \times\ 2.6 \\ \hline \end{array}$

4) $\begin{array}{r} 21.09 \\ \times\ 9.07 \\ \hline \end{array}$

7) $\begin{array}{r} 6.9 \\ \times\ 0.8 \\ \hline \end{array}$

2) $\begin{array}{r} 8.7 \\ \times\ 6.9 \\ \hline \end{array}$

5) $\begin{array}{r} 14.3 \\ \times\ 15.7 \\ \hline \end{array}$

8) $\begin{array}{r} 67.08 \\ \times\ 10 \\ \hline \end{array}$

3) $\begin{array}{r} 4.06 \\ \times\ 7.05 \\ \hline \end{array}$

6) $\begin{array}{r} 7.09 \\ \times\ 4.0 \\ \hline \end{array}$

9) $\begin{array}{r} 12.08 \\ \times\ 1000 \\ \hline \end{array}$

✎ **Find each quotient.**

10) $18.7 \div 2.5$

13) $8.05 \div 2.5$

16) $7.38 \div 1000$

11) $45.2 \div 5$

14) $7.6 \div 100$

17) $36.1 \div 100$

12) $15.6 \div 4.5$

15) $2.07 \div 10$

18) $0.03 \div 10$

Converting Between Fractions, Decimals and Mixed Numbers

✍ **Convert fractions to decimals**

1) $\dfrac{5}{10}$ 4) $\dfrac{15}{16}$ 7) $\dfrac{32}{40}$

2) $\dfrac{36}{100}$ 5) $\dfrac{6}{18}$ 8) $\dfrac{14}{25}$

3) $\dfrac{5}{8}$ 6) $\dfrac{40}{100}$ 9) $\dfrac{73}{10}$

✍ **Convert decimal into fraction or mixed numbers.**

10) 0.7   14) 0.16   18) 0.06

11) 7.25   15) 0.05   19) 0.68

12) 0.44   16) 0.18   20) 4.6

13) 5.3   17) 0.4   21) 3.2

Answers of Worksheets – Chapter 2

Simplifying Fractions

1) $\frac{21}{31}$

2) $\frac{3}{4}$

3) $\frac{2}{3}$

4) $\frac{3}{4}$

5) $\frac{1}{3}$

6) $\frac{1}{4}$

7) $\frac{4}{9}$

8) $\frac{1}{2}$

9) $\frac{2}{5}$

10) $\frac{1}{9}$

11) $\frac{5}{9}$

12) $\frac{3}{4}$

13) $5\frac{5}{8}$

14) $2\frac{13}{16}$

15) $9\frac{1}{5}$

16) $3\frac{5}{8}$

17) $1\frac{1}{2}$

18) $4\frac{1}{5}$

19) $1\frac{3}{4}$

20) $\frac{9}{10}$

21) $7\frac{1}{2}$

Factoring Numbers

1) $1, 2, 4, 8, 16, 32$

2) $1, 2, 13, 26$

3) $1, 2, 4, 5, 8, 10, 20, 40$

4) $1, 2, 3, 4, 6, 8, 12\ 7, 16, 24, 32, 48, 96$

5) $1, 2, 3, 4, 5, 6, 10, 12, 15, 20, 30, 60$

6) $1, 2, 4, 7, 14, 28$

7) $1, 5, 7, 35$

8) $1, 2, 4, 8, 16, 32, 64$

9) $1, 2, 4, 7, 8, 14, 56$

10) $1, 3, 5, 15, 25, 75$

11) $1, 3, 9, 27, 81$

12) $1, 2, 3, 4, 6, 8, 12, 16, 24, 48$

13) $2 \times 2 \times 5$

14) 5×13

15) $3 \times 3 \times 11$

16) $3 \times 2 \times 7$

17) $2 \times 2 \times 3 \times 7$

18) $3 \times 2 \times 5$

19) $2 \times 2 \times 13$

20) $3 \times 5 \times 7$

21) $3 \times 5 \times 5$

22) $2 \times 2 \times 2 \times 2 \times 3$

23) $2 \times 2 \times 3 \times 3$

24) 31×1

25) $2 \times 2 \times 2 \times 3$

26) $2 \times 3 \times 3 \times 3$

Greatest Common Factor

1) 7	7) 4	13) 6
2) 1	8) 15	14) 15
3) 12	9) 36	15) 21
4) 5	10) 9	16) 25
5) 13	11) 16	17) 2
6) 19	12) 5	18) 31

Least Common Multiple

1) 18	7) 126	13) 96
2) 175	8) 84	14) 525
3) 192	9) 704	15) 168
4) 36	10) 60	16) 315
5) 42	11) 150	17) 336
6) 45	12) 40	18) 360

Divisibility Rules

1) 24   <u>2</u> <u>3</u> <u>4</u> 5 <u>6</u> 7 <u>8</u> 9 10
2) 32   <u>2</u> 3 <u>4</u> 5 6 7 <u>8</u> 9 10
3) 16   <u>2</u> 3 <u>4</u> 5 6 7 <u>8</u> 9 10
4) 42   <u>2</u> <u>3</u> 4 5 <u>6</u> <u>7</u> 8 9 10
5) 28   <u>2</u> 3 <u>4</u> 5 6 <u>7</u> 8 9 10
6) 56   <u>2</u> 3 <u>4</u> 5 6 <u>7</u> <u>8</u> 9 10
7) 48   <u>2</u> 3 <u>4</u> 5 <u>6</u> 7 <u>8</u> 9 10
8) 36   <u>2</u> <u>3</u> <u>4</u> 5 <u>6</u> 7 8 <u>9</u> 10
9) 81   2 <u>3</u> 4 5 6 7 8 <u>9</u> 10
10) 50   <u>2</u> 3 4 <u>5</u> 6 7 8 9 <u>10</u>
11) 63   2 <u>3</u> 4 5 6 <u>7</u> 8 <u>9</u> 10
12) 84   <u>2</u> 3 <u>4</u> 5 <u>6</u> 7 8 9 10

Adding and Subtracting Fractions

1) $\frac{5}{6}$

2) $\frac{20}{21}$

3) $\frac{55}{63}$

4) $\frac{3}{4}$

5) $\frac{7}{10}$

6) $\frac{3}{5}$

7) 1

8) $\frac{23}{26}$

9) $\frac{53}{112}$

10) $\frac{1}{2}$

11) $\frac{9}{40}$

12) $\frac{23}{42}$

13) $\frac{1}{72}$

14) $\frac{11}{60}$

15) $\frac{5}{16}$

16) $\frac{1}{75}$

17) $\frac{1}{36}$

18) $\frac{5}{112}$

Multiplying and Dividing Fractions

1) $\frac{3}{28}$

2) $\frac{1}{10}$

3) $\frac{1}{12}$

4) $\frac{4}{11}$

5) $\frac{3}{5}$

6) $\frac{36}{85}$

7) 0

8) 1

9) $\frac{1}{6}$

10) 0

11) $\frac{1}{12}$

12) $\frac{8}{11}$

13) $\frac{3}{4}$

14) $\frac{1}{2}$

15) $\frac{2}{5}$

16) $\frac{30}{7}$

17) $\frac{4}{5}$

18) $\frac{13}{17}$

19) 54

20) 1

21) $\frac{84}{85}$

Adding Mixed Numbers

1) $5\frac{1}{2}$

2) 8

3) $3\frac{1}{3}$

4) $5\frac{5}{6}$

5) $10\frac{17}{30}$

6) $5\frac{10}{21}$

7) $2\frac{5}{12}$

8) $4\frac{29}{35}$

9) $7\frac{11}{24}$

10) $12\frac{1}{8}$

11) $2\frac{35}{72}$

12) $5\frac{15}{16}$

Subtract Mixed Numbers

1) $3\frac{1}{4}$

2) $3\frac{1}{4}$

3) $\frac{5}{14}$

4) $4\frac{7}{36}$

5) $\frac{2}{3}$

6) $4\frac{5}{8}$

7) 3

8) $2\frac{19}{33}$

9) $\frac{31}{35}$

10) $2\frac{1}{6}$

11) $2\frac{1}{3}$

12) $4\frac{2}{35}$

Multiplying Mixed Numbers

1) $1\frac{9}{16}$

2) $4\frac{1}{21}$

3) $5\frac{10}{21}$

4) $5\frac{31}{40}$

5) $8\frac{1}{15}$

6) $1\frac{23}{27}$

7) $6\frac{3}{16}$

8) $9\frac{9}{10}$

9) $8\frac{2}{3}$

10) 12

11) $1\frac{5}{9}$

12) $2\frac{4}{5}$

Dividing Mixed Numbers

1) $1\frac{5}{8}$

2) $2\frac{1}{2}$

3) 3

4) $1\frac{3}{7}$

5) $\frac{20}{21}$

6) $1\frac{1}{16}$

7) $\frac{5}{6}$

8) $1\frac{1}{5}$

9) $1\frac{19}{56}$

10) $\frac{5}{12}$

11) $2\frac{11}{98}$

12) 0

Comparing Decimals

1) <

2) >

3) =

4) <

5) >

6) <

7) >

8) <

9) >

10) >

11) <

12) <

13) <

14) >

15) >

16) =

17) <

18) <

19) >

20) >

Rounding Decimals

1) 1.0

2) 3.0

3) 7.6

4) 0.4

5) 8

6) 0.09

7) 6.2

8) 76.76

9) 3.63

10) 12.4	16) 37	22) 57.1
11) 1.1	17) 420	23) 126.5
12) 4.3	18) 312.7	24) 7.01
13) 3.2	19) 18.4	25) 0.01
14) 6.05	20) 86	26) 11.34
15) 94	21) 3.21	27) 6.4

Adding and Subtracting Decimals

1) 22.11	5) 200.23	9) 5.65
2) 80	6) 11.61	10) 6.853
3) 31.83	7) 1.78	11) 1.01
4) 40.65	8) 9.55	12) 38.34

Multiplying and Dividing Decimals

1) 14.3	7) 5.52	13) 3.22
2) 60.03	8) 670.8	14) 0.207
3) 28.623	9) 12080	15) 0.076
4) 191.2863	10) 7.48	16) 0.00738
5) 224.51	11) 9.04	17) 0.361
6) 28.36	12) 3.46...	18) 0.003

Converting Between Fractions, Decimals and Mixed Numbers

1) 0.5	9) 7.3	16) $\frac{9}{50}$
2) 0.36	10) $\frac{7}{10}$	17) $\frac{2}{5}$
3) 0.625	11) $7\frac{1}{4}$	18) $\frac{3}{50}$
4) 0.9375	12) $\frac{11}{25}$	19) $\frac{17}{25}$
5) 0.333...	13) $5\frac{3}{10}$	20) $4\frac{3}{5}$
6) 0.4	14) $\frac{4}{25}$	21) $3\frac{1}{5}$
7) 0.8	15) $\frac{1}{20}$	
8) 0.56		

Chapter 3: Proportion, Ratio, Percent

Topics that you'll learn in this chapter:

- ✓ Writing and Simplifying Ratios

- ✓ Create a Proportion

- ✓ Similar Figures

- ✓ Simple Interest

- ✓ Ratio and Rates Word Problems

- ✓ Percentage Calculations

- ✓ Converting Between Percent, Fractions, and Decimals

- ✓ Percent Problems

- ✓ Markup, Discount, and Tax

"Do not worry about your difficulties in mathematics. I can assure you mine are still greater." – Albert Einstein

Unlock problems

❖ Writing Ratios

✓ A ratio is a comparison of two numbers. Ratio can be written as a division

Example: $3:5$, or $\frac{3}{8}$ and $\frac{5}{8}$.

❖ Simplifying Ratios

✓ You can calculate equivalent ratios by multiplying or dividing both sides of the ratio by the same number.

Examples:

$3:6 = 1:2$

$4:9 = 8:18$

❖ Create a Proportion

✓ A proportion contains 2 equal fractions! A proportion simply means that two fractions are equal.

Example: $\frac{2}{4} = \frac{8}{16}$

❖ Similar Figures

✓ Two or more figures are similar if the corresponding angles are equal, and the corresponding sides are in proportion.

Example:

3–4–5 triangle is similar to a 6–8–10 triangle

❖ Ratio and Rates Word Problems

✓ To solve a ratio or a rate word problem, create a proportion and use cross multiplication method!

Example:

$$\frac{x}{4} = \frac{8}{16}$$

$$16x = 4 \times 8 \Longrightarrow x = 2$$

❖ Percentage Calculations

✓ Use the following formula to find part, whole, or percent:

$$\text{part} = \frac{\text{percent}}{100} \times \text{whole}$$

Example:

$$\frac{20}{100} \times 100 = 20$$

❖ Percent Problems

✓ Base = Part ÷ Percent

✓ Part = Percent × Base

✓ Percent = Part ÷ Base

Example:

2 is 10% of 20.

$$2 \div 0.10 = 20$$

$$2 = 0.10 \times 20$$

$$0.10 = 2 \div 20$$

❖ Markup, Discount, and Tax

Markup = selling price − cost

✓ Markup rate = markup divided by the cost

Discount:

✓ Multiply the regular price by the rate of discount

✓ Selling price =original price − discount

Tax:

✓ To find tax, multiply the tax rate to the taxable amount (income, property value, etc.)

Example:

✓ Original price of a microphone: $49.99, discount: 5%, tax: 5%

Selling price = 49.87

❖ Simple Interest

Simple Interest: The charge for borrowing money or the return for lending it.

✓ Interest = principal x rate x time

$$I = prt$$

Example:

$450 at 7% for 8 years.

$I = prt$

$I = 450 \times 0.07 \times 8 = \$252$

❖ Converting Between Percent, Fractions, and Decimals

✓ To a percent: Move the decimal point 2 places to the right and add the % symbol.

✓ Divide by 100 to convert a number from percent to decimal.

Examples:

$30\% = 0.30$

$0.24 = 24\%$

Writing and Simplifying Ratios

✍ **Express each ratio as a rate and unite rate.**

1) 75 dollars for 5 chairs.

2) 203miles on 7 gallons of gas.

3) 168 miles on 3 hours

4) 16 inches of snow in 24 hours

5) 42 dimes t0 126 dimes

6) 27 feet out of 81 feet

✍ **Express each ratio as a fraction in the simplest form.**

7) 17 cups to 51 cups

8) 24 cakes out of 60 cakes

9) 42 red desks out of 189 desks

10) 45 story books out of 72 books

11) 28 gallons to 40 gallons

12) 68 miles out of 100 miles

✍ **Reduce each ratio.**

1) 24: 42

2) 45: 15

3) 28: 36

4) 24: 26

5) 14: 56

6) 48: 60

7) 108: 252

8) 81: 45

9) 100: 25

10) 18: 32

11) 60: 10

12) 35: 45

13) 76: 57

14) 10: 100

15) 16: 40

16) 17: 34

17) 5: 25

18) 66: 39

Create a Proportion

✍ **Create proportion from the given set of numbers**

1) 2, 1, 8, 4

2) 7, 24, 56, 3

3) 21, 18, 63, 6

4) 11, 15, 22, 30

5) 5, 30, 75, 2

6) 9, 7, 54, 42

7) 32, 2, 16, 4

8) 63, 12, 9, 84

9) 10, 10, 100, 1

Similar Figures

✍ **Each pair of figures is similar. Find the missing side.**

1)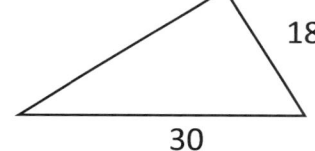

2)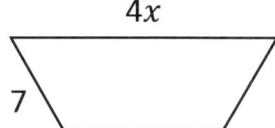

3)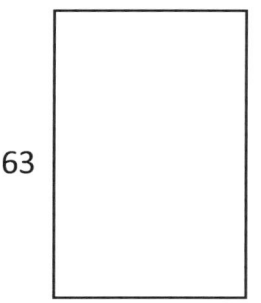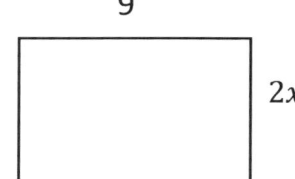

Ratio and Rates Word Problems

✎ Solve.

1) In Peter's class, 27 of the students are tall and 15 are short. In Elise's class 81 students are tall and 45 students are short. Which class has a higher ratio of tall to short students?

2) In a party, 12 soft drinks are required for every 26 guests. If there are 364 guests, how many soft drinks is required?

3) The price of 6 bananas at the first Market is $1.08. The price of 4 of the same bananas at second Market is $0.76. Which place is the better buy?

4) You can buy 4 cans of green beans at a supermarket for $2.40. How much does it cost to buy 32 cans of green beans?

5) The bakers at a Bakery can make 132 bagels in 6 hours. How many bagels can they bake in 8 hours? What is that rate per hour?

Percentage Calculations

Calculate the percentages.

1) 25% of 38

2) 42% of 8

3) 15% of 15

4) 63% of 75

5) 4% of 50

6) 35% of 14

7) 18% of 3

8) 9% of 47

9) 10% of 100

10) 50% of 72

11) 75% of 60

12) 95% of 12

13) 80% of 30

14) 11% of 120

15) 1% of 210

16) 32% of 0

Solve.

17) What percentage of 40 is 2

18) 13.2 is what percentage of 88?

19) 38 is what percentage of 76?

20) Find what percentage of 85 is 22.1.

Percent Problems

✍ **Solve each problem.**

1) 64% of what number is 16?   5) 18 is what percent of 20?

2) What is 70% of 140 inches?   6) 34 is 40% of what?

3) What percent of 58 is 23.2?   7) 9 is what percent of 12?

4) 8 is 250% of what?   8) 95% of 100 is what number?

9) Mia require 60% to pass. If she gets 240 marks and falls short by 60 marks, what were the maximum marks she could have got?

10) Jack scored 34 out of 40 marks in mathematics, 8 out of 10 marks in history and 78 out of 100 marks in science. In which subject his percentage of marks is the best?

Markup, Discount, and Tax

✍ **Find the selling price of each item.**

1) Cost of a chair: $18.99, markup: 25%, discount: 8%, tax: 8%

2) Cost of computer: $1,490.00, markup: 60%

3) Cost of a pen: $2.50, markup: 60%, discount: 20%, tax: 5%

4) Cost of a puppy: $1,900, markup: 38%, discount: 15%

Simple Interest

✎ **Use simple interest to find the ending balance.**

1) $3,200 at 13.7% for 2 years.

2) $280,000 at 3.75% for 15 years.

3) $2,000 at 1.9% for 5 years.

4) $14,700 at 5.8% for 3 years.

5) $47,500 at 0.5% for 16 months.

6) Emily puts $4,500 into an investment yielding 2.75% annual simple interest; she left the money in for six years. How much interest does Sara get at the end of those six years?

7) A new car, valued at $36,000, depreciates at 8.5% per year from original price. Find the value of the car 5 years after purchase.

8) $360 interest is earned on a principal of $1,800 at a simple interest rate of 5% interest per year. For how many years was the principal invested?

Converting Between Percent, Fractions, and Decimals

✎ Converting fractions to decimals

1) $\dfrac{40}{100}$

2) $\dfrac{28}{100}$

3) $\dfrac{4}{25}$

4) $\dfrac{1}{10}$

5) $\dfrac{7}{20}$

6) $\dfrac{2}{100}$

7) $\dfrac{40}{50}$

8) $\dfrac{25}{10}$

9) $\dfrac{6}{30}$

✎ Write each decimal as a percent.

10) 0.25

11) 1.2

12) 0.015

13) 0.005

14) 0.725

15) 0.2

16) 1.05

17) 0.0275

18) 0.0025

19) 0.175

Answers of Worksheets – Chapter 3

Writing Ratios

1) $\frac{75 \text{ dollars}}{5 \text{ books}}$, 15.00 dollars per chair

2) $\frac{203 \text{ miles}}{7 \text{ gallons}}$, 29 miles per gallon

3) $\frac{168 \text{ miles}}{3 \text{ hours}}$, 56 miles per hour

4) $\frac{120" \text{ of snow}}{24 \text{ hours}}$, 5 inches of snow per hour

5) $\frac{126 \text{ dimes}}{42 \text{ dimes}}$, 3 per dime

6) $\frac{108 \text{ feet}}{27 \text{ feet}}$, 4 per foot

7) $\frac{1}{3}$

8) $\frac{2}{5}$

9) $\frac{2}{9}$

10) $\frac{5}{8}$

11) $\frac{7}{10}$

12) $\frac{17}{25}$

Reduce each Ratio

1) 4: 7

2) 3: 1

3) 7: 9

4) 12: 13

5) 1: 4

6) 4: 5

7) 3: 7

8) 9: 5

9) 4: 1

10) 9: 16

11) 6: 1

12) 7: 9

13) 4: 3

14) 1: 10

15) 2: 5

16) 1: 2

17) 1: 5

18) 22: 13

Create a Proportion

1) 1: 4 = 2: 8

2) 3: 24 = 7: 56

3) 6: 18 = 21: 63

4) 11: 22 = 15: 30

5) 5: 75=2: 30

6) 7: 42 =9: 54

7) 2: 16 =4: 32

8) 9: 63 =12: 84

9) 1: 10 =10: 100

Similar Figures

1) 5

2) 2

3) 2

Ratio and Rates Word Problems

1) The ratio for both classes is equal to 9 to 5.

2) 168

3) The price at the first Market is a better buy.

4) $19.20

5) 176, the rate is 22 per hour.

Percentage Calculations

1) 9.5	6) 4.9	11) 45	16) 0
2) 3.36	7) 0.54	12) 11.4	17) 5%
3) 2.25	8) 4.23	13) 24	18) 15%
4) 47.25	9) 10	14) 13.2	19) 50%
5) 2	10) 36	15) 2.1	20) 26%

Percent Problems

1) 25	4) 3.2	7) 75%	10) Mathematics
2) 98	5) 90%	8) 95	
3) 40%	6) 85	9) 500	

Markup, Discount, and Tax

1) $23.59	3) $3.36
2) $2,384	4) $2,228.70

Simple Interest

1) $4,076.80	4) $17,257.80	7) $20,700
2) $437,500.00	5) $51,300.00	8) 4 years
3) $2,190.00	6) $742.50	

Converting Between Percent, Fractions, and Decimals

1) 0.4	6) 0.02	11) 120%	16) 105%
2) 0.28	7) 0.8	12) 1.5%	17) 2.75%
3) 0.16	8) 2.5	13) 0.5%	18) 0.25%
4) 0.1	9) 0.2	14) 72.5%	19) 17.5%
5) 0.35	10) 25%	15) 20%	

Chapter 4: Exponents and Radicals

Topics that you'll learn in this chapter:

✓ Multiplication Property of Exponents

✓ Division Property of Exponents

✓ Powers of Products and Quotients

✓ Zero, Negative Exponents and Bases

Mathematics is no more computation than typing is literature.

– John Allen Paulos

Unlock problems

❖ Multiplication Property of Exponents

Exponents rules

- ✓ $x^a . x^b = x^{a+b}$

- ✓ $\dfrac{x^a}{x^b} = x^{a-b}$

- ✓ $\dfrac{1}{x^b} = x^{-b}$

- ✓ $(x^a)^b = x^{a \times b}$

- ✓ $(xy)^a = x^a . y^a$

Example:

$$(x^2 y)^3 = x^6 y^3$$

❖ Division Property of Exponents

- ✓ $\dfrac{x^a}{x^b} = x^{a-b} \; ; x \neq 0$

Example:

$$\dfrac{x^{12}}{x^5} = x^7$$

❖ Powers of Products and Quotients

- ✓ For any nonzero numbers a and b and any integer

$$(ab)^x = a^x b^x$$

Example:

$$(2x^2 2 . y^3)^2 = 4x^2 . y^6$$

❖ Zero and Negative Exponents

✓ A negative exponent simply means that the base is on the wrong side of the fraction line, so you need to flip the base to the other side. For instance, "x^{-2}" (pronounced as "ecks to the minus two") just means "x^2" but underneath, as in $\frac{1}{x^2}$

Example:

$$-5^{-2} = \frac{1}{25}$$

❖ Negative Exponents and Negative Bases

✓ Make the power positive. A negative exponent is the reciprocal of that number with a positive exponent.

✓ The parenthesis is important!

-5^{-2} is not the same as $(-5)^{-2}$

$-5^{-2} = -\frac{1}{5^2}$ and $(-5)^{-2} = \frac{1}{(-5)^2} = +\frac{1}{5^2}$

Example:

$$2x^{-3} = \frac{2}{x^3}$$

❖ Writing Scientific Notation

✓ It is used to write very big or very small numbers in decimal form.

✓ In scientific notation all numbers are written in the form of:

$m \times 10^n$

Decimal notation	Scientific notation
5	5×10^0
$-25{,}000$	-2.5×10^4
0.5	5×10^{-1}
2,122.456	2.1×10^3

❖ Square Roots

✓ A square root of x is a number r whose square is: $r^2 = x$

✓ r is a square root of x.

Example:

$$\sqrt{4} = 2$$

❖ Simplifying Radical Expressions

For square roots:

✓ Step 1: Find the prime factors of the numbers inside the radical.

✓ Step 2: Find the largest perfect score factor of the number.

✓ Step 3: Rewrite the radical as the product of perfect score and its matching factor and simplify.

❖ Simplifying Radical Expressions Involving Fractions

✓ Radical expressions cannot be in the denominator. (number in the bottom)

✓ To get rid of the radical in the denominator, multiply

both numerator and denominator by the radical in the denominator.

✓ If there is a radical and another integer in the denominator, multiply both numerator and denominator by the conjugate of the denominator.

✓ The conjugate of $a + b$ is $a - b$ and vice versa.

Example:

$$\frac{3}{\sqrt{5} - 3} \times \frac{\sqrt{5} + 3}{\sqrt{5} + 3} = \frac{3(\sqrt{5} + 3)}{\sqrt{5}^2 - 3^2} = \frac{3\sqrt{5} + 9}{5 - 9} = \frac{3\sqrt{5} + 9}{-4}$$

❖ Multiplying Radical Expressions

To multiply radical expressions:

✓ Step 1: Multiply the numbers outside of the radicals.

✓ Step 2: Multiply the numbers inside the radicals.

✓ Step 3: Simplify if needed.

Example:

$$2\sqrt{3} \times 4\sqrt{2} = 2 \times 4 \times \sqrt{3 \times 2} = 8\sqrt{6}$$

❖ Adding and Subtracting Radical Expressions

✓ Only numbers that have the same radical part can be added or subtracted.

✓ Remember, combining "unlike" radical terms is not possible.

✓ For number with the same radical part, just add or subtract factors outside the radicals.

Example:

$$4\sqrt{2} + 6\sqrt{2} = 10\sqrt{2}$$

❖ Domain and Range of Radical Functions

- ✓ To find domain and rage of radical functions, remember that having a negative number under the square root symbol is not possible. (for square roots)
- ✓ To find the domain of the function, find all possible values of the variable inside radical.
- ✓ To find the range, plugin the minimum and maximum values of the variable inside radical.

❖ Solving Radical Equations

- ✓ Step 1: Isolate the radical on one side of the equation.
- ✓ Step 2: Square both sides of the equation to remove the radical
- ✓ Step 3: Solve the equation for the variable
- ✓ Step 4: Plugin the answer into the original equation to avoid extraneous values.

Example:

$$2\sqrt{x + 2} = 10 \rightarrow \sqrt{x + 2} = 5 \rightarrow (\sqrt{x + 2})^2 = 5^2 \rightarrow x + 2 = 25 \rightarrow x = 23$$

$$2\sqrt{x + 2} = 10 \rightarrow 2\sqrt{23 + 2} = 10 \rightarrow 2\sqrt{25} = 10 \rightarrow 2 \times 5 = 10!$$

Multiplication Property of Exponents

✎ Simplify.

1) $3^2 \times 3^2$

2) $4 . 4^2 . 4^2$

3) $2^2 . 2^2$

4) $5x^3 . x$

5) $14x^4 . 2x$

6) $5x . 2x^2$

7) $6x^4 . 7x^4$

8) $4x^2 . 6x^3 y^4$

9) $8x^2 y^5 . 8xy^3$

10) $5xy^4 . 4x^3 y^3$

11) $(3x^2)^2$

12) $4x^5 y^3 . 5x^2 y^3$

13) $7x^3 . 10y^3 x^5 . 7yx^3$

14) $(x^4)^3$

15) $(3x^2)^4$

16) $8x^4 y^5 . 2x^2 y^3$

Division Property of Exponents

✎ Simplify.

1) $\dfrac{5^6}{5}$

2) $\dfrac{43}{43^{45}}$

3) $\dfrac{3^2}{3^3}$

4) $\dfrac{5^4}{5^2}$

5) $\dfrac{x}{x^{13}}$

6) $\dfrac{24x^3}{6x^4}$

7) $\dfrac{2x^{-5}}{11x^{-2}}$

8) $\dfrac{49x^8}{7x^3}$

9) $\dfrac{11x^6}{4x^7}$

10) $\dfrac{42x^2}{4x^3}$

11) $\dfrac{x}{10x^3}$

12) $\dfrac{x^3}{2x^5}$

13) $\dfrac{16x^3}{14x^6}$

14) $\dfrac{12x^3}{6y^8}$

15) $\dfrac{25xy^4}{x^6y^2}$

16) $\dfrac{2x^4}{7x}$

17) $\dfrac{32x^2y^8}{4x^3}$

18) $\dfrac{12x^4}{15x^7y^9}$

19) $\dfrac{yx^4}{10yx^8}$

20) $\dfrac{16x^4y}{9x^8y^2}$

21) $\dfrac{6x^8}{36x^8}$

Powers of Products and Quotients

✎ Simplify.

1) $(x^3)^4$

2) $(2xy^4)^2$

3) $(6x^4)^2$

4) $(12x^5)^2$

5) $(2x^2y^4)^4$

6) $(3x^4y^4)^3$

7) $(4x^2y^2)^2$

8) $(5x^4y^3)^4$

9) $(4x^6y^8)^2$

10) $(15x^3 . x)^3$

11) $(x^9 \ x^6)^3$

12) $(7x^{10}y^3)^3$

13) $(6x^3 \ x^2)^2$

14) $(4x^3 \ 5x)^2$

15) $(10x^{11}y^3)^2$

16) $(8x^7 \ y \ ^5)^2$

17) $(9x^4y^6)^5$

18) $(3x^4)^2$

19) $(3x \ 4y^3)^2$

20) $(7x^2y)^3$

21) $(14x^2y^5)^2$

Zero and Negative Exponents

✎ **Evaluate the following expressions.**

1) 5^{-2}

2) 3^4

3) 7^{-2}

4) 5^{-4}

5) 12^{-1}

6) 7^{-1}

7) 6^{-2}

8) 8^{-2}

9) 5^{-2}

10) 15^{-1}

11) 7^{-3}

12) 0^5

13) 10^{-7}

14) 4^{-4}

15) 4^{-2}

16) 2^{-3}

17) 3^{-4}

18) 6^{-1}

19) 7^3

20) 11^{-2}

21) $\left(\frac{3}{4}\right)^{-2}$

22) $\left(\frac{1}{5}\right)^{-2}$

23) $\left(\frac{1}{2}\right)^{-6}$

24) $\left(\frac{2}{5}\right)^{-2}$

25) 10^{-4}

26) 1^{-100}

Negative Exponents and Negative Bases

✎ **Simplify.**

1) -4^{-1}

2) $-5x^{-3}$

3) $\frac{x}{x^{-3}}$

4) $-\frac{a^{-6}}{b^{-2}}$

5) $\frac{5}{x^{-3}}$

6) $\frac{b}{-9c^{-4}}$

7) $-\frac{25n^{-2}}{10p^{-3}}$

8) $\frac{4ab^{-2}}{-3c^{-2}}$

9) $10x^2y^{-3}$

10) $\left(-\frac{1}{4}\right)^{-2}$

11) $\left(-\frac{5}{4}\right)^{-2}$

12) $\left(\frac{x}{3yz}\right)^{-3}$

Writing Scientific Notation

✎ **Write each number in scientific notation.**

1) 81×10^5

2) 50

3) 0.0000008

4) 254000

5) 0.000225

6) 6.5

7) 0.00063

8) 89000000

9) 9000000

10) 85000000

11) 0.0000036

12) 0.00015

13) 0.008

14) 8600

15) 1960

16) 170000

17) 0.115

18) 0.05

Square Roots

✎ **Find the value each square root.**

1) $\sqrt{81}$

2) $\sqrt{0}$

3) $\sqrt{36}$

4) $\sqrt{64}$

5) $\sqrt{49}$

6) $\sqrt{1}$

7) $\sqrt{25}$

8) $\sqrt{9}$

9) $\sqrt{144}$

10) $\sqrt{121}$

11) $\sqrt{16}$

12) $\sqrt{256}$

13) $\sqrt{100}$

14) $\sqrt{169}$

15) $\sqrt{324}$

16) $\sqrt{90}$

17) $\sqrt{484}$

18) $\sqrt{529}$

Simplifying Radical Expressions

✍ *Simplify.*

1) $\sqrt{33x^2}$

2) $\sqrt{40x^2}$

3) $\sqrt{25x^3}$

4) $\sqrt{144a}$

5) $\sqrt{512v}$

6) $\sqrt{9x^2}$

7) $\sqrt{384}$

8) $\sqrt{162p^3}$

9) $\sqrt{125m^4}$

10) $\sqrt{693x^3y^3}$

11) $\sqrt{81x^3y^3}$

12) $\sqrt{9a^4b^3}$

13) $\sqrt{40x^3y^3}$

14) $3\sqrt{45x^2}$

15) $5\sqrt{60x^2}$

16) $4\sqrt{81a}$

17) $3\sqrt{8x^2y^3r}$

18) $4\sqrt{64x^2y^3z^4}$

Simplifying Radical Expressions Involving Fractions

✎ *Simplify.*

1) $\dfrac{2\sqrt{7r}}{\sqrt{m^4}}$

7) $\dfrac{\sqrt{6}-\sqrt{4}}{\sqrt{4}-\sqrt{6}}$

2) $\dfrac{6\sqrt{2}}{\sqrt{k}}$

8) $\dfrac{\sqrt{3}}{\sqrt{7}-2}$

3) $\dfrac{\sqrt{c}}{\sqrt{c}+\sqrt{d}}$

9) $\dfrac{\sqrt{2}-\sqrt{6}}{\sqrt{2}+\sqrt{6}}$

4) $\dfrac{5+\sqrt{3}}{2-\sqrt{3}}$

10) $\dfrac{3\sqrt{5}+5}{2\sqrt{5}-3}$

5) $\dfrac{2+\sqrt{7}}{6-\sqrt{5}}$

11) $\dfrac{\sqrt{8a^5b^3}}{\sqrt{2ab^2}}$

6) $\dfrac{3}{2+\sqrt{3}}$

12) $\dfrac{6\sqrt{20x^3}}{3\sqrt{5x}}$

Multiplying Radical Expressions

✎ *Simplify.*

1) $\sqrt{12x} \times \sqrt{12x}$

2) $-5\sqrt{27} \times -3\sqrt{3}$

3) $3\sqrt{45x^2} \times \sqrt{5x^2}$

4) $\sqrt{8x^2} \times \sqrt{12x^3}$

5) $-10\sqrt{8} \times \sqrt{5x^3}$

6) $5\sqrt{21} \times \sqrt{3}$

7) $\sqrt{3} \times -\sqrt{64}$

8) $-5\sqrt{16x^3} \times 4\sqrt{2x^2}$

9) $\sqrt{12}\,(2 + \sqrt{3})$

10) $-3\sqrt{8}\,(2 + \sqrt{8})$

11) $\sqrt{12x}\,(3 - \sqrt{6x})$

12) $\sqrt{3x}\,(x^3 + \sqrt{27})$

13) $\sqrt{15r}\,(2 + \sqrt{3})$

14) $\sqrt{2v}\,(\sqrt{6} + \sqrt{10})$

15) $(-2\sqrt{6} + 3)\,(\sqrt{6} - 1)$

16) $(2 - \sqrt{3})(-2 + \sqrt{3})$

17) $(10 - 4\sqrt{5})(6 - \sqrt{5})$

18) $(\sqrt{6} - \sqrt{3})(\sqrt{6} + \sqrt{3})$

Adding and Subtracting Radical Expressions

✐*Simplify.*

1) $6\sqrt{10} + 4\sqrt{10}$

2) $-3\sqrt{12} - 3\sqrt{27}$

3) $-3\sqrt{22} - 5\sqrt{22}$

4) $-9\sqrt{7} + 12\sqrt{7}$

5) $6\sqrt{3} - \sqrt{27}$

6) $-\sqrt{18} + 4\sqrt{2}$

7) $-4\sqrt{7} + 4\sqrt{7}$

8) $3\sqrt{27} + 3\sqrt{3}$

9) $2\sqrt{20} - 2\sqrt{5}$

10) $3\sqrt{18} - \sqrt{2}$

11) $-10\sqrt{35} + 14\sqrt{35}$

12) $-4\sqrt{19} - 5\sqrt{19}$

13) $-3\sqrt{6x} - 3\sqrt{6x}$

14) $\sqrt{5y^2} + y\sqrt{20}$

15) $\sqrt{8mn^2} + n\sqrt{18m}$

16) $-8\sqrt{27a} - 2\sqrt{3a}$

17) $-6\sqrt{7ab} - 6\sqrt{7ab}$

18) $\sqrt{27a^2b} + a\sqrt{12b}$

Solving Radical Equations

✎ **Solve each equation. Remember to check for extraneous solutions.**

1) $\sqrt{x-6} = 3$

2) $2 = \sqrt{x-3}$

3) $\sqrt{r} = 5$

4) $\sqrt{m+8} = 4$

5) $5\sqrt{3x} = 15$

6) $1 = \sqrt{x-4}$

7) $-18 = -6\sqrt{r+3}$

8) $10 = 2\sqrt{35v}$

9) $\sqrt{n+3} - 1 = 6$

10) $\sqrt{3r} = \sqrt{2r-2}$

11) $\sqrt{3x+15} = \sqrt{x+5}$

12) $\sqrt{v} = \sqrt{2v-5}$

13) $\sqrt{12-x} = \sqrt{x-2}$

14) $\sqrt{m+5} = \sqrt{3m+5}$

15) $\sqrt{2r+20} = \sqrt{-16-2r}$

16) $\sqrt{k+5} = \sqrt{1-k}$

17) $-10\sqrt{x-10} = -50$

18) $\sqrt{36-x} = \sqrt{\dfrac{x}{5}}$

Domain and Range of Radical Functions

✍️ *Identify the domain and range of each.*

1) $y = \sqrt{x + 4} - 3$

2) $y = \sqrt[3]{x - 3} + 6$

3) $y = \sqrt{x - 2} - 2$

4) $y = \sqrt[3]{x + 1} - 5$

Sketch the graph of each function.

5) $y = \sqrt{x} + 2$

6) $y = 3\sqrt{-x} - 2$

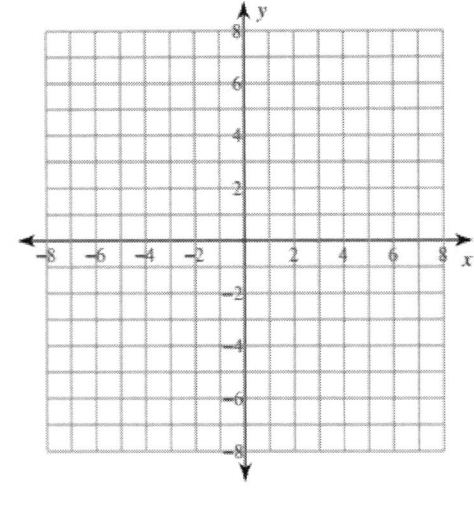

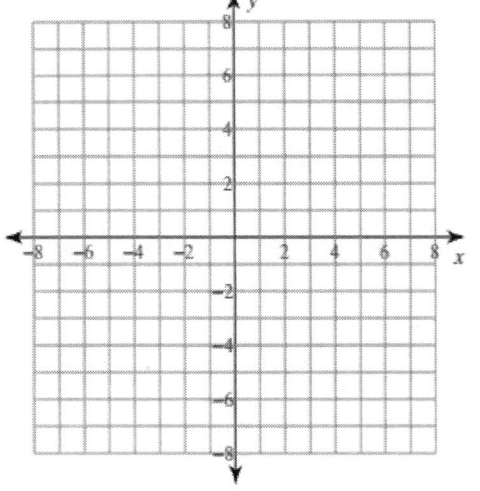

Answers of Worksheets – Chapter 4

Multiplication Property of Exponents

1) 3^4

2) 4^5

3) 2^4

4) $5x^4$

5) $48x^5$

6) $10x^3$

7) $42x^8$

8) $24x^5y^4$

9) $64x^3y^8$

10) $20x^4y^7$

11) $9x^4$

12) $20x^7y^6$

13) $490x^{11}y^4$

14) x^{12}

15) $81x^8$

16) $16x^6y^8$

Division Property of Exponents

1) 5^5

2) $\dfrac{1}{43^{44}}$

3) $\dfrac{1}{3}$

4) 5^2

5) $\dfrac{1}{x^{12}}$

6) $\dfrac{4}{x}$

7) $\dfrac{2}{11x^3}$

8) $7x^5$

9) $\dfrac{11}{4x}$

10) $\dfrac{21}{2x}$

11) $\dfrac{1}{10x^2}$

12) $\dfrac{1}{2x^2}$

13) $\dfrac{8}{7x^3}$

14) $\dfrac{2x^3}{y^8}$

15) $\dfrac{25y^2}{x^5}$

16) $\dfrac{2x^3}{7}$

17) $\dfrac{8y^8}{x}$

18) $\dfrac{4}{5x^3y^9}$

19) $\dfrac{1}{10x^4}$

20) $\dfrac{16}{9x^4y}$

21) $\dfrac{1}{6}$

Powers of Products and Quotients

1) x^{12}

2) $4x^2y^8$

3) $36x^8$

4) $124x^{10}$

5) $16x^8y^{16}$

6) $27x^{12}y^{12}$

7) $16x^4y^4$

8) $625x^{16}y^{12}$

9) $16x^{12}y^{16}$

10) $3,375x^{12}$

11) x^{45}

12) $343x^{30}y^9$

13) $36x^{10}$

14) $350x^8$

15) $100x^{22}y^6$

16) $64x^{14}y^{10}$

17) $59,049x^{20}y^{30}$

18) $9x^8$

19) $144x^2y^6$

20) $343x^6y^3$

21) $196x^4y^{10}$

Zero and Negative Exponents

1) $\frac{1}{25}$

2) 81

3) $\frac{1}{49}$

4) $\frac{1}{625}$

5) $\frac{1}{12}$

6) $\frac{1}{7}$

7) $\frac{1}{36}$

8) $\frac{1}{64}$

9) $\frac{1}{25}$

10) $\frac{1}{15}$

11) $\frac{1}{343}$

12) 0

13) $\frac{1}{10000000}$

14) $\frac{1}{256}$

15) $\frac{1}{16}$

16) $\frac{1}{8}$

17) $\frac{1}{81}$

18) $\frac{1}{6}$

19) 343

20) $\frac{1}{121}$

21) $\frac{16}{9}$

22) 25

23) 64

24) $\frac{25}{4}$

25) $\frac{1}{10000}$

26) 1

Negative Exponents and Negative Bases

1) $-\frac{1}{4}$

2) $-\frac{5}{x^3}$

3) x^4

4) $-\frac{b^2}{a^6}$

5) $5x^3$

6) $-\frac{bc^4}{9}$

7) $-\frac{5p^3}{2n^2}$

8) $-\frac{4ac^2}{3b^2}$

9) $\frac{10x^2}{y^3}$

10) 16

11) $\frac{16}{25}$

12) $\frac{27y^3z^3}{x^3}$

Writing Scientific Notation

1) 8.1×10^6

2) 5×10^1

3) 8×10^{-7}

4) 2.54×10^5

5) 2.25×10^{-4}

6) 6.5×10^0

7) 6.3×10^{-4}

8) 8.9×10^7

9) 9×10^6

10) 8.5×10^7

11) 3.6×10^{-6}

12) 1.5×10^{-4}

13) 8×10^{-3}

14) 8.6×10^3

15) 1.96×10^3

16) 1.7×10^5

17) 1.15×10^{-1}

18) 5×10^{-2}

Square Roots

1) 9

2) 0

3) 6

4) 8

5) 7

6) 1

7) 5

8) 3

9) 12

10) 11

11) 4

12) 16

13) 10

14) 13

15) 18

16) 30

17) 22   18) 23

Simplifying radical expressions

1) $\sqrt{33}\,x$

2) $2x\sqrt{10}$

3) $5x\,\sqrt{x}$

4) $12\,\sqrt{a}$

5) $8\,\sqrt{8v}$

6) $3x$

7) $8\,\sqrt{6}$

8) $9p\,\sqrt{2p}$

9) $5m^2\,\sqrt{5}$

10) $3x.\,y\,\sqrt{77xy}$

11) $9x.\,y\,\sqrt{xy}$

12) $3a^2.\,b\,\sqrt{b}$

13) $2x.\,y\,\sqrt{10xy}$

14) $9x\,\sqrt{5}$

15) $10x\,\sqrt{15}$

16) $36\,\sqrt{a}$

17) $6x\,y\,\sqrt{2yr}$

18) $32z^2.\,x.\,y\,\sqrt{y}$

Simplifying radical expressions involving fractions

1) $\frac{2\sqrt{7r}}{m^2}$

2) $\frac{6\sqrt{2k}}{k}$

3) $\frac{c-\sqrt{cd}}{c-d}$

4) $13+7\,\sqrt{3}$

5) $\frac{2\sqrt{7}+6\sqrt{5}}{31}$

6) $6-3\,\sqrt{3}$

7) -1

8) $\frac{\sqrt{21}+2\sqrt{3}}{3}$

9) $2+\sqrt{3}$

10) $\frac{19\sqrt{5}+45}{11}$

11) $2a^2\,\sqrt{b}$

12) $4x$

Multiplying radical expressions

1) $12x$

2) 135

3) $45x^2$

4) $4x^{2.}\,\sqrt{6x}$

5) $-20x^{2.}\,\sqrt{10}$

6) $15\sqrt{7}$

7) $-8\,\sqrt{3}$

8) $-80x^2\,\sqrt{2x}$

9) $4\,\sqrt{3}+6$

10) $-(12\,\sqrt{2}+24)$

11) $6\,\sqrt{3x}-6x\,\sqrt{2}$

12) $\sqrt{3}.\,x^2+9x$

13) $3\sqrt{5r}+2\sqrt{15r}$

14) $2\,\sqrt{3v}+2\sqrt{5v}$

15) $5\sqrt{6}-15$

16) $4\sqrt{3}-7$

17) $80-34\sqrt{5}$

18) 3

Adding and subtracting radical expressions

1) $10\sqrt{10}$

2) $-15\sqrt{3}$

3) $-8\sqrt{22}$

4) $3\sqrt{7}$

5) $3\sqrt{3}$

6) $\sqrt{2}$

7) 0

8) $12\sqrt{3}$

9) $2\sqrt{5}$

10) $8\sqrt{2}$

11) $4\sqrt{35}$

12) $-9\sqrt{19}$

13) $-6\sqrt{6x}$

14) $3y\sqrt{5}$

15) $5n\sqrt{2m}$

16) $-26\sqrt{3}\,a$

17) $-12\sqrt{7ab}$

18) $5a\sqrt{3b}$

Solving radical equations

1) {15}

2) {7}

3) {25}

4) {8}

5) {3}

6) {5}

7) {0}

8) $\{\frac{5}{7}\}$

9) {46}

10) {-2}

11) {−5}

12) {5}

13) {7}

14) {0}

15) {−9}

16) {−2}

17) {35}

18) {30}

Domain and range of radical functions

1) domain: $x \geq -4$

   range: $y \geq -3$

2) domain: {all real numbers}

   range: {all real numbers}

3) domain: $x \geq 3$

   range: $y \geq 6$

4) domain: {all real numbers}

   range: {all real numbers}

5)

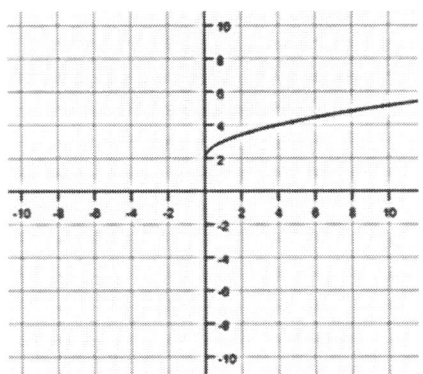

6)

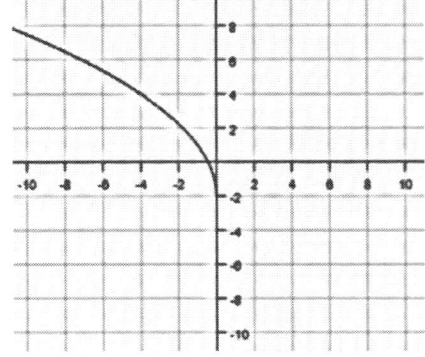

Chapter 5: Sequences and Series

Topics that you'll learn in this chapter:

- ✓ Arithmetic Sequences
- ✓ Geometric Sequences
- ✓ Comparing Arithmetic and Geometric Sequences
- ✓ Finite Geometric Series
- ✓ Infinite Geometric Series

Mathematics is like checkers in being suitable for the young, not too difficult, amusing, and without peril to the state. – Plato

Unlock problems

❖ Arithmetic Sequences

✓ $x_n = a + d(n - 1)$

a = the first term

d = the common difference between terms

n = how many terms to add up

❖ Geometric Sequences

✓ $x_n = ar^{(n-1)}$

a = the first term

r = the common ratio

❖ Comparing Arithmetic and Geometric Sequences

✓ Arithmetic Sequence: There is a constant difference between consecutive terms. (each term is the sum of previous term and a constant.

✓ Geometric Sequence: The consecutive terms are in a constant ratio. (each term is the product of previous term and a ratio)

❖ Finite Geometric Series

✓ Finite Geometric Series: The sum of a geometric series is finite when the absolute value of the ratio is less than 1.

✓ $Sn = \sum_{i=1}^{n} ar^{i-1} = a_1(\frac{1-r^n}{1-r})$

❖ Infinite Geometric Series

✓ Infinite Geometric Series: The sum of a geometric series is infinite when the absolute value of the ratio is more than 1.

✓ $S = \sum_{i=0}^{\infty} a_i r^i = \frac{a_1}{1-r}$

Arithmetic Sequences

✍️Given the first term and the common difference of an arithmetic sequence find the first five terms and the explicit formula.

1) $a_1 = 23$, $d = 2$

3) $a_1 = 15$, $d = 10$

2) $a_1 = -10$, $d = -2$

4) $a_1 = -30$, $d = -50$

✍️Given a term in an arithmetic sequence and the common difference find the first five terms and the explicit formula.

5) $a_{36} = -248$, $d = -6$

7) $a_{38} = -52.3$, $d = -1.1$

6) $a_{34} = 156$, $d = 5$

8) $a_{20} = -591$, $d = -30$

✍️Given a term in an arithmetic sequence and the common difference find the recursive formula and the three terms in the sequence after the last one given.

9) $a_{22} = -46$, $d = -2$

11) $a_{18} = 26.4$, $d = 1.1$

10) $a_{12} = 28.6$, $d = 1.2$

12) $a_{32} = -1.2$, $d = 0.6$

Geometric Sequences

✍ Determine if the sequence is geometric. If it is, find the common ratio.

1) -1, 5, -25, 125, …

3) 3, 16, 23, 64, …

2) $-3, -9, -27, -81, …$

4) $-2, -8, -16, -32, …$

✍ Given the first term and the common ratio of a geometric sequence find the first five terms and the explicit formula.

5) $a_1 = 0.6$, $r = -5$

6) $a_1 = 1$, $r = 3$

✍ Given the recursive formula for a geometric sequence find the common ratio, the first five terms, and the explicit formula.

7) $a_n = a_{n-1} \cdot 2$, $a_1 = 3$

9) $a_n = a_{n-1} \cdot 5$, $a_1 = 1$

8) $a_n = a_{n-1} \cdot -2$, $a_1 = -2$

10) $a_n = a_{n-1} \cdot \frac{1}{3}$, $a_1 = -3$

✍ Given two terms in a geometric sequence find the 8th term and the recursive formula.

11) $a_4 = 216$ and $a_5 = -1296$

12) $a_5 = -32$ and $a_2 = -4$

Comparing Arithmetic and Geometric Sequences

✍ **For each sequence, state if it arithmetic, geometric, or neither.**

1) 1, 3, 6, 9, 12, …

2) 1, 4, 16, 64, 256, …

3) 4, 24, 64, 100, …

4) −28, −30, −32, −34, −36, …

5) −5, 15, −45, 135, −405, …

6) 40, 43, 46, 49, 52, …

7) 1, 4, 7, 10, 13, …

8) −34, −27, −20, −13, −6, …

9) $a_n = -145 + 200_n$

10) $a_n = 16 + 3_n$

11) $a_n = -2 . (-3)^{n-1}$

12) $a_n = -23 + 4_n$

13) $a_n = (3n)^2$

14) $a_n = -40 + 7_n$

15) $a_n = -(-2)^{n-1}$

16) $a_n = 3 . (-3)^{n-1}$

Finite Geometric Series

✎ Evaluate the related series of each sequence.

1) $1, -6, 36, -216$

3) $7, -21, 61, -183$

2) $4, 8, 16, 32, 64$

4) $-2, 1, -\frac{1}{2}, \frac{1}{4}, -\frac{1}{8}$

✎ Evaluate each geometric series described.

5) $1 + 5 + 25 + 125 \dots, n = 6$

13) $\sum_{p=1}^{6} (-3) \cdot (-4)^{p-1}$

6) $1 - 3 + 9 - 27 \dots, n = 9$

14) $\sum_{m=1}^{9} (-3)^{m-1} 6$

7) $-2 - 8 - 32 - 128 \dots, n = 9$

15) $\sum_{n=1}^{9} 3^{n-1} 6$

8) $4 - 8 + 16 - 32 \dots, n = 7$

9) $1 + \frac{1}{2} + \frac{1}{4} + \frac{1}{8} \dots, n = 6$

16) $\sum_{n=1}^{4} (\frac{1}{3})^{n-1} 4$

10) $-5 - 5 - 5 - 5 \dots, n = 12$

17) $\sum_{n=1}^{6} 4^{n-1} + \sum_{j=1}^{7} (-3)^{j-1}$

11) $\sum_{n=1}^{8} 2 \cdot (-3)^{n-1}$

18) $5 \sum_{n=1}^{8} 5^{n-1}$

12) $\sum_{n=1}^{10} 5 \cdot 2^{n-1}$

Infinite Geometric Series

✍ **Determine if each geometric series converges or diverges.**

1) $a_1 = -1$, $r = 3$

2) $a_1 = -3$, $r = 4$

3) $a_1 = 5.5$, $r = 0.5$

4) $81 + 27 + 9 + 3 \ldots,$

5) $-3 + \dfrac{12}{5} - \dfrac{48}{25} + \dfrac{192}{125} \ldots,$

6) $\dfrac{128}{3125} - \dfrac{64}{625} + \dfrac{32}{125} - \dfrac{16}{25} \ldots,$

✍ **Evaluate each infinite geometric series described.**

7) $a_1 = 3$, $r = -\dfrac{1}{5}$

8) $a_1 = 1$, $r = -4$

9) $a_1 = 1$, $r = -3$

10) $a_1 = 3$, $r = \dfrac{1}{2}$

11) $1 + 0.5 + 0.25 + 0.125 + \ldots$

12) $1 - 0.6 + 0.36 - 0.216 \ldots,$

13) $81 - 27 + 9 - 3 \ldots,$

14) $3 + \dfrac{9}{4} + \dfrac{27}{16} + \dfrac{81}{64} \ldots,$

15) $\sum_{k=1}^{\infty} 4^{k-1}$

16) $\sum_{i=1}^{\infty} \left(\dfrac{1}{3}\right)^{i-1}$

Answers of Worksheets – Chapter 5

Arithmetic Sequences

1) First Five Terms: 23, 25, 27, 29, 31, Explicit: $a_n = 23 + 2(n-1)$

2) First Five Terms: −10, −12, −14, −16, −18, Explicit: $a_n = -10 - 2(n-1)$

3) First Five Terms: 15, 25, 35, 45, 55, Explicit: $a_n = 15 + 10(n-1)$

4) First Five Terms: −30, −80, −130, −180, −230,

   Explicit: $a_n = -30 - 50(n-1)$

5) First Five Terms: −38, −44, −50, −56, −62, Explicit: $a_n = -38 - 6(n-1)$

6) First Five Terms: −9, −4, 1, 6, 11, Explicit: $a_n = -9 + 5(n-1)$

7) First Five Terms: −11.6, −12.7, −13.8, −14.9, −16,

   Explicit: $a_n = -11.6 - 1.1(n-1)$

8) First Five Terms: −21, −51, −81, −111, −141,

   Explicit: $a_n = -21 - 30(n-1)$

9) Next 3 terms: −48, −50, −52, Recursive: $a_n = a_{n-1} - 2$, $a_1 = -4$

10) Next 3 terms: 29.8, 31, 32.2, Recursive: $a_n = a_{n-1} + 1.2$, $a_1 = 15.4$

11) Next 3 terms: 27.5, 28.6, 29.7, Recursive: $a_n = a_{n-1} + 1.1$, $a_1 = 7.7$

12) Next 3 terms: −0.6, 0, 0.6, Recursive: $a_n = a_{n-1} + 0.6$, $a_1 = -19.8$

Geometric Sequences

1) $r = -5$ 2) $r = 3$ 3) not geometric 4) not geometric

5) First Five Terms: 0.6, −3, 15, −75, 375; Explicit: $a_n = 0.6 \cdot (-5)^{n-1}$

6) First Five Terms: 1, 3, 9, 27, 81; Explicit: $a_n = 3^{n-1}$

7) Common Ratio: $r = 2$; First Five Terms: 3, 6, 12, 24. 48

   Explicit: $a_n = 3 \cdot 2^{n-1}$

8) Common Ratio: $r = -2$; First Five Terms: −2, 4, −8, 16, −32

   Explicit: $an = -2 \cdot (-2)^{n-1}$

9) Common Ratio: $r = 5$; First Five Terms: 1, 5, 25, 125, 625

Explicit: $a_n = 1.5^{n-1}$

10) Common Ratio: r = 2; First Five Terms: $-3, -1, -\frac{1}{3}, -\frac{1}{9}, -\frac{1}{27}$

Explicit: $an = -3.\left(\frac{1}{3}\right)^{n-1}$

11) $a_8 = -279936$, Recursive: $a_n = a_{n-1}. -6$, $a_1 = 1$

12) $a_8 = -256$, Recursive: $a_n = a_{n-1}. 2$, $a_1 = -2$

Comparing Arithmetic and Geometric Sequences

1) Neither	7) Neither	13) Neither
2) Geometric	8) Arithmetic	14) Arithmetic
3) Neither	9) Arithmetic	15) Geometric
4) Arithmetic	10) Arithmetic	16) Geometric
5) Geometric	11) Geometric	
6) Arithmetic	12) Arithmetic	

Finite Geometric

1) -185	7) $-174,762$	13) 2,457
2) 124	8) 172	14) 29,526
3) -140	9) $\frac{63}{32}$	15) 59,046
4) $-\frac{11}{8}$	10) -60	16) $\frac{160}{27}$
5) 3,906	11) $-3,280$	17) 1,912
6) 4,921	12) 5,115	18) 488,280

Infinite Geometric

1) Diverges	7) $\frac{5}{2}$	12) 0.625
2) Diverges	8) No sum	13) $\frac{243}{4}$
3) Converges	9) No sum	14) 12
4) Converges	10) 6	15) No sum
5) Converges	11) 2	16) $\frac{3}{2}$
6) Diverges		

Chapter 6: Algebraic Expressions

Topics that you'll learn in this chapter:

✓ Expressions and Variables

✓ Simplifying Variable and Polynomial Expressions

✓ Translate Phrases into an Algebraic Statement

✓ The Distributive Property

✓ Evaluating One and two Variable

✓ Combining like Terms

Without mathematics, there's nothing you can do. Everything around you are mathematics. Everything around you are numbers." – Shakuntala Devi

Unlock problems

❖ Expressions and Variables

✓ A variable is a letter that represents unknown numbers. A variable can be used in the same manner as all other numbers:

Addition	$2 + a$	2 plus a
Subtraction	$y - 3$	y minus 3
Division	$\dfrac{4}{x}$	4 divided by x
Multiplication	$5a$	5 times a

❖ Translate Phrases into an Algebraic Statement

Translating key words and phrases into algebraic expressions:

✓ **Addition**: plus, more than, the sum of, etc.

✓ **Subtraction**: minus, less than, decreased, etc.

✓ **Multiplication**: times, product, multiplied, etc.

✓ **Division**: quotient, divided, ratio, etc.

Example:

eight more than a number is 20

$8 + x = 20$

❖ The Distributive Property

✓ Distributive Property:

$$a\,(b + c) = ab + ac$$

Example:

$$3(4 + 3x) = 12 + 9x$$

❖ Simplifying Variable Expressions

✓ Combine "like" terms. (values with same variable and same power)

✓ Use distributive property if necessary.

Distributive Property: $a(b + c) = ab + ac$

Example:

$$2x + 2(1 - 5x) = 2x + 2 - 10x = -8x + 2$$

❖ Simplifying Polynomial Expressions

✓ In mathematics, a polynomial is an expression consisting of variables and coefficients that involves only the operations of addition, subtraction, multiplication, and non–negative integer exponents of variables.

$$P(x) = a_0x^n + a_1x^{n-1} + \ ... \ + a_{n-2}x^2 + a_{n-1}x + a_n$$

Example:

An example of a polynomial of a single indeterminate x is $x^2 - 4x + 7$.

An example for three variables is

$$x^3 + 2xyz^2 - yz + 1$$

❖ Evaluating One Variable

✓ To evaluate one variable expression, find the variable and substitute a number for that variable.

✓ Perform the arithmetic operations.

Example:

$4x + 8, x = 6$

$4(6) + 8 = 24 + 8 = 32$

❖ Evaluating Two Variables

✓ To evaluate an algebraic expression, substitute a number for each variable and perform the arithmetic operations.

Example:

$2x + 4y - 3 + 2,$

$x = 5, y = 3$

$2(5) + 4(3) - 3 + 2 = 10 + 12 - 3 + 2 = 21$

❖ Combining like Terms

✓ Terms are separated by "+" and "−" signs.

✓ Like terms are terms with same variables and same powers.

✓ Be sure to use the "+" or "−" that is in front of the coefficient.

Example:

$22x + 6 + 2x = 24x + 6$

Translate Phrases into an Algebraic Statement

✎ **Write an algebraic expression for each phrase.**

1) Ten subtracted from a number.

2) The quotient of fifteen and a number.

3) A number increased by forty.

4) A number divided by $- 10$.

5) The difference between fifty–two and a number.

6) Twice a number decreased by 35.

7) five times the sum of a number and $- 12$.

8) The quotient of 50 and the product of a number and $- 6$.

The Distributive Property

✎ **Use the distributive property to simply each expression.**

1) $6(8 - 2x)$

2) $-(-4 - 2x)$

3) $(-6x + 3)(-1)$

4) $(-4)(x - 2)$

5) $2(10 + 2x)$

6) $(-6x + 8)2$

7) $(3 - 6x)(-2)$

8) $(-12)(x + 1)$

9) $(-4)(x - 2) + 3(x + 1)$

10) $(-x)(-1 + 4x) - 4x(2 + 3x)$

11) $3(-5x - 1) + 4(2 - 3x)$

12) $(-2)(x + 2) - (4 + 3x)$

Evaluating One Variable

✎ Simplify each algebraic expression.

1) $3x + 2, x = 2$

2) $x + (-4), x = -2$

3) $-2x - 5, x = -1$

4) $\left(-\frac{32}{x}\right) - 9 + x, x = 4$

5) $\frac{35}{x} - 3, x = 7$

6) $(-2) + \frac{x}{4} + 4x, x = 8$

7) $10 + 3x - 2, x = -2$

8) $(-2) + \frac{x}{7}, x = 49$

9) $\left(-\frac{18}{x}\right) - 9 + 2x, x = 2$

10) $(-2) + \frac{3x}{8}, x = 32$

Evaluating Two Variables

✎ Simplify each algebraic expression.

1) $6a - (6 - b),$

   $a = 3, b = 2$

2) $4x + 2y - 4 + 2,$

   $x = 2, y = 3$

3) $\left(-\frac{18}{x}\right) + 1 + 5y,$

   $x = 9, y = 2$

4) $(-2)(-2a - 2b),$

   $a = 4, b = 2$

5) $6x + 2 - 2y,$

   $x = 6, y = 8$

6) $8 + 2(-2x - 3y),$

   $x = 1, y = 4$

7) $10x + y,$

   $x = 6, y = 9$

8) $x \times 4 \div y,$

   $x = 3, y = 2$

Expressions and Variables

✎ Simplify each expression.

1) $5(-2 - 9x), x = 2$

2) $-2(6 - 6x) - 3x, x = 1$

3) $x - 6x, x = 5$

4) $2x + 10x, x = 4$

5) $9 - 3x + 9x + 3, x = 1$

6) $5(3x + 2), x = 3$

7) $2(3 - 2x) - 6, x = 2$

8) $5x + 2x - 8, x = 2$

9) $3x + 6y, \ x = 5, y = 3$

10) $2x \ + \ 5x, x = 4,$

11) $5(-2x + 9) + 4, x = 7,$

12) $8x - 2x - 5, x = 2,$

13) $2x - 3x - 9, x = 5$

14) $(-4)\,(-2x - 8y), x = 3, y = 1$

15) $8x + 3 + 6\,y, x = 4, y = 3$

16) $(-6)(-8x - 9y), x = 3, y = 3$

Combining like Terms

✎ **Simplify each expression.**

1) $-6(-3x + 2)$

2) $3(-4 + 2x)$

3) $-4 - 4x + 6x + 9$

4) $6x - 3x - 5 + 8$

5) $(-3)(6x - 2) + 18$

6) $3(3x + 4) + 8x$

7) $2(-4x - 7) + 4(3x + 2)$

8) $(2x - 3y)3 + 5y$

9) $2.5x^2 \times (-6x)$

10) $-3 - x^2 - 8x^2$

11) $6 + 10x^2 + 2$

12) $9(-2x - 6) + 8$

13) $6x^2 + 4x + 5x^2$

14) $4x^2 - 10x^2 + 16x$

15) $4x^2 - 6x - x$

16) $(-4)(5x - 3)$

17) $3x + 6(2 - 2x)$

18) $10x + 3(18x - 4)$

19) $2(x + 8)$

20) $-3(x + 1) + 2x$

21) $x - 6y + 6x + 3y - 4x$

22) $7(-2x + 2y) + 10x - 8y$

23) $(-2x) - 7 + 15x + 3$

24) $3(x + 1) + 12(x - 1)$

Simplifying Polynomial Expressions

✎ Simplify each polynomial.

1) $(x^2 + 1) - (4 + x^2)$

2) $(23x^3 - 10x^2) - (4x^2 - 9x^3)$

3) $4x^5 - 5x^6 + 5x^5 - 16x^6 + 3x^6$

4) $(-3x^5 + 9 - 4x) + (8x^4 + 5x + 5x^5)$

5) $10x^2 - 5x^4 + 12x^3 - 20x^4 + 10x^3$

6) $-6x^2 + 5x^2 + 7x^3 + 16 + 32$

7) $5x^3 + 1 + 4x^2 - 3x - 10x$

8) $14x^2 - 6x^3 - 2x(2x^2 + x)$

9) $(2x^4 - x) - (2x - x^4)$

10) $(10x^3 + 2x^4) - (3x^4 - x^3)$

11) $(12 + 2x^3) + (4x^3 + 4)$

12) $(5x^2 - 3) + (x^2 - 3x^3)$

Answers of Worksheets – Chapter 6

Translate Phrases into an Algebraic Statement

1) $x - 10$

2) $15/x$

3) $x + 40$

4) $\dfrac{x}{-10}$

5) $52 - x$

6) $2x - 35$

7) $5(x + (-12))$

8) $\dfrac{50}{-6x}$

The Distributive Property

1) $-12x + 48$

2) $2x + 4$

3) $6x - 3$

4) $-4x + 8$

5) $4x + 20$

6) $-12x + 16$

7) $12x - 6$

8) $-12x - 12$

9) $-x + 11$

10) $-16x^2 - 7x$

11) $-27x + 5$

12) $-5x - 8$

Evaluating One Variable

1) 8

2) -6

3) -3

4) -13

5) 2

6) 32

7) 2

8) 5

9) -14

10) 10

Evaluating Two Variables

1) 14

2) 12

3) 9

4) 24

5) 22

6) -20

7) 69

8) 6

Expressions and Variables

1) -100

2) -3

3) -25

4) 48

5) 18

6) 55

7) -8

8) 6

9) 33

10) 28

11) -21

12) 7

13) -14

14) 56

15) 53

16) 306

Combining like Terms

1) $18x - 12$

2) $6x-12$

3) $5 + 2x$

4) $3x + 3$

5) $24 - 18x$

6) $17x + 12$

7) $4x - 6$

8) $6x - 4y$

9) $-15x^3$

10) $-9x^2 - 3$

11) $10x^2 + 8$

12) $-18x - 48$

13) $11x^2 + 4x$

14) $-6x^2 + 16x$

15) $4x^2 - 7x$

16) $-20x + 12$

17) $-9x + 12$

18) $64x - 12$

19) $2x + 16$

20) $-x - 3$

21) $3x - 3y$

22) $4x + 6y$

23) $13x - 4$

24) $15x - 9$

Simplifying Polynomial Expressions

1) -3

2) $32x^3 - 14x^2$

3) $-18x^6 + 9x^5$

4) $2x^5 + 8x^4 + x + 9$

5) $-25x^4 + 22x^3 + 10x^2$

6) $7x^3 - x^2 + 48$

7) $5x^3 + 4x^2 - 13x + 1$

8) $-10x^3 + 12x^2$

9) $3x^4 - 3x$

10) $-x^4 + 11x^3$

11) $6x^3 + 16$

12) $-3x^3 + 6x^2 - 3$

Chapter 7: Equations and Inequalities

Topics that you'll learn in this chapter:

✓ One, Two, and Multi – Step Equations

✓ Graphing Single– Variable Inequalities

✓ One, Two, and Multi – Step Inequalities

✓ Solving Systems of Equations by Substitution and Elimination

✓ Finding Slope and Writing Linear Equations

✓ Graphing Lines Using Slope– Intercept and Standard Form

✓ Graphing Linear Inequalities

✓ Finding Midpoint and Distance of Two Points

"The study of mathematics, like the Nile, begins in minuteness but ends in magnificence." – Charles Caleb Colton

Unlock problems

❖ One–Step Equations

✓ The values of two expressions on both sides of an equation are equal.

$$ax + b = c$$

✓ You only need to perform one Math operation in order to solve the equation.

Example:

$$-8x = 16 \implies x = -2$$

❖ Two–Step Equations

✓ You only need to perform two math operations (add, subtract, multiply, or divide) to solve the equation.

✓ Simplify using the inverse of addition or subtraction.

✓ Simplify further by using the inverse of multiplication or division.

Example:

$$-2(x - 1) = 42$$
$$(x - 1) = -21$$
$$x = -20$$

❖ Multi–Step Equations

✓ Combine "like" terms on one side.

✓ Bring variables to one side by adding or subtracting.

✓ Simplify using the inverse of addition or subtraction.

✓ Simplify further by using the inverse of multiplication or division.

Example:

$$3x + 15 = -2x + 5$$

Add $2x$ both sides

$$5x + 15 = +5$$

Subtract 15 both sides

$$5x = -10$$

Divide by 5 both sides

$$x = -2$$

❖ Graphing Single–Variable Inequalities

✓ Isolate the variable.

✓ Find the value of the inequality on the number line.

✓ For less than or greater than draw open circle on the value of the variable.

✓ If there is an equal sign too, then use filled circle.

✓ Draw a line to the right direction.

❖ One–Step Inequalities

✓ Isolate the variable.

✓ For dividing both sides by negative numbers, flip the direction of the inequality sign.

Example:

$$x + 4 \geq 11 \ \rightarrow x \geq 7$$

❖ Two–Step Inequalities

✓ Isolate the variable.

✓ For dividing both sides by negative numbers, flip the direction of the of the inequality sign.

✓ Simplify using the inverse of addition or subtraction.

✓ Simplify further by using the inverse of multiplication or division.

Example: $2x + 9 \geq 11 \Longrightarrow 2x \geq 2 \Longrightarrow x \geq 1$

❖ Multi–Step Inequalities

✓ Isolate the variable.

✓ Simplify using the inverse of addition or subtraction.

✓ Simplify further by using the inverse of multiplication or division.

Example: $\frac{7x + 1}{3} \geq 5 \Longrightarrow 7x + 1 \geq 15 \Longrightarrow 7x \geq 14 \Longrightarrow x \geq 7$

❖ Solving Systems of Equations by Substitution

✓ Consider the system of equations

$$x + y = 1, -2x + y = 4$$

Substitute $x = 1 - y$ in the second equation

$$-2(1 - y) + y = 4 \rightarrow -2 + 2y + y = 4 \Longrightarrow y = 2$$

Substitute $y = 2$ in $x = 1 - y$

$x = 1 - 2 = -1 ; (-1,2)$

Example:

$-2x - 2y = -13$

$-4x + 2y = 10$

$(0.5, 6)$

❖ Solving Systems of Equations by Elimination

✓ The elimination method for solving systems of linear equations uses the addition property of equality. You can add the same value to each side of an equation.

Example:

$$x + 2y = 6$$
$$+ \ -x + y = 3$$
$$\overline{}$$

$3y = 9 \implies y = 3$

$x + 2y = 6 \implies x + 6 = 6 \implies x = 0$

❖ Systems of Equations Word Problems

✓ Define your variables, write two equations, and use one of the methods for solving systems of equations to solve.

❖ Graphing Lines Using Standard Form

✓ Find the −intercept of the line by putting zero for y.

✓ Find the y−intercept of the line by putting zero for the x.

✓ Connect these two points.

Example:

$x + 4y = 12$

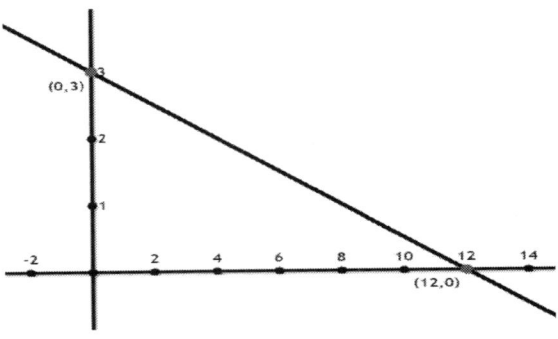

❖ Graphing Linear Inequalities

✓ First, graph the "equals" line.

✓ Choose a testing point. (it can be any point on both sides of the line.)

✓ Put the value of (x, y) of that point in the inequality. If that works, that part of the line is the solution. If the values don't work, then the other part of the line is the solution.

❖ Finding Midpoint

✓ Midpoint of the segment AB: $M(\frac{x_1+x_2}{2}, \frac{y_1+y_2}{2})$

Example: $(3, 9), (- 1, 6) \Rightarrow M(1, 7.5)$

❖ Finding Distance of Two Points

✓ Distance from A to B: $d = \sqrt{(x_1 - x_2)^2 + (y_1 - y_2)^2}$

Example: $(- 1, 2), (- 1, - 7) \Rightarrow d = 9$

One–Step Equations

✎ **Solve each equation.**

1) $x + 6 = 16$

2) $42 = (-6) + x$

3) $5x = (-50)$

4) $(-81) = (-9x)$

5) $(-6) = 4 + x$

6) $3 + x = (-4)$

7) $10x = (-110)$

8) $12 = x + 6$

9) $(-25) + x = (-20)$

10) $6x = (-36)$

11) $x - 18 = (-20)$

12) $x - 9 = (-24)$

13) $(-30) = x - 25$

14) $(-7x) = 49$

15) $(-66) = (6x)$

16) $x - 10 = 30$

17) $6x = 30$

18) $36 = (-9x)$

19) $2x = 68$

20) $25x = 500$

Two–Step Equations

✎ **Solve each equation.**

1) $3(2 + x) = 9$

2) $(-6)(x - 3) = 42$

3) $(-10)(2x - 3) = (-10)$

4) $4(1 + x) = -12$

5) $14(2x + 1) = 42$

6) $6(3x + 2) = 42$

7) $3(7 + 2x) = (-60)$

8) $(-10)(2x - 3) = 48$

9) $2(x + 5) = 30$

10) $\frac{2x - 6}{4} = 2$

11) $(-24) = \frac{x + 3}{6}$

12) $110 = (-5)(x - 2)$

13) $\frac{x}{5} - 8 = 2$

14) $-15 = 9 + \frac{x}{6}$

15) $\frac{12 + x}{4} = (-10)$

16) $(-2)(6 + 2x) = (-100)$

17) $(-5x) + 10 = 30$

18) $\frac{x + 6}{5} = (-5)$

19) $\frac{x + 36}{5} = (-5)$

20) $(-8) + \frac{x}{4} = (-12)$

Multi–Step Equations

✍ **Solve each equation.**

1) $-(3-2x)=7$

2) $-18=-(3x+12)$

3) $5x-15=(-x)+3$

4) $-225=(-3x)-12x$

5) $4(1+2x)+2x=-16$

6) $4x-10=3+x-5+x$

7) $10-2x=(-32)-2x+2x$

8) $7-3x-3x=3-3x$

9) $26+11x+x=(-30)+4$

10) $(-3x)-8(-1+5x)=352$

11) $36=(-6x)-2+2$

12) $35=2x-14+5x$

13) $5(1+5x)=-495$

14) $-40=(-4x)-6x$

15) $x+5=(-7)+5x$

16) $5x-8=8x+4$

17) $10=-(x-8)$

18) $(-18)-6x=6(1+3x)$

19) $x+2=-3(6+3x)$

20) $5=1-2x+4$

Graphing Single–Variable Inequalities

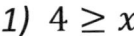 Draw a graph for each inequality.

1) $4 \geq x$

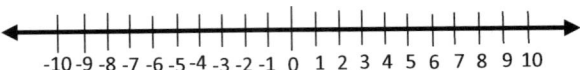

2) $x < -2$

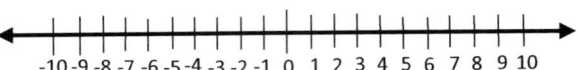

3) $-3 < x$

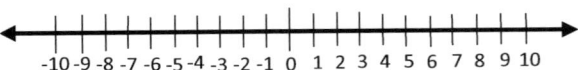

4) $-x \geq 1$

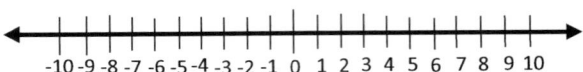

5) $x > 2$

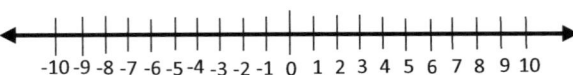

6) $-0.5 \leq x$

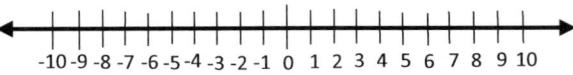

One–Step Inequalities

Solve each inequality and graph it.

1) $x + 3 \geq 9$

2) $x - 7 \leq 4$

3) $-4x < 2$

4) $-x + 5 > -8$

5) $x + 5 \geq -11$

6) $6x < 12$

7) $5x > -20$

Two–Step Inequalities

✍ Solve each inequality and graph it.

1) $3x - 4 \leq 5$

2) $2x - 2 \leq 6$

3) $4x - 4 \leq 8$

4) $3x + 6 \geq 12$

5) $6x - 5 \geq 19$

6) $2x - 4 \leq 6$

7) $8x - 4 \leq 4$

8) $6x + 4 \leq 10$

9) $5x + 4 \leq 9$

10) $7x - 4 \leq 3$

11) $4x - 19 < 19$

12) $2x - 3 < 21$

13) $7 + 4x \geq 19$

14) $9 + 4x < 21$

15) $3 + 2x \geq 19$

16) $6 + 4x < 22$

Multi–Step Inequalities

✍ Solve each inequality.

1) $\frac{9x}{7} - 7 < 2$

2) $\frac{4x + 8}{2} \leq 12$

3) $\frac{3x - 8}{7} > 1$

4) $-3(x - 7) > 21$

5) $4 + \frac{x}{3} < 7$

6) $\frac{2x + 6}{4} \leq 10$

Solving Systems of Equations by Substitution

Solve each system of equation by substitution.

1) $-3x + 3y = 3$

$$x + y = 3$$

2) $-10x + 2y = -6$

$$3x - 8y = 24$$

3) $y = -6$

$$15x - 10y = 75$$

4) $2y = -6x + 10$

$$10x - 8y = -6$$

5) $3x - 2y = 5$

$$3y = 3x - 3$$

6) $2x + 3y = 5$

$$3x + y = -3$$

7) $x + 10y = 6$

$$x + 5y = 1$$

8) $2x + 4y = 16$

$$x - 4, y = -1$$

Solving Systems of Equations by Elimination

✎ Solve each system of equation by elimination.

1) $-5x + y = -5$

$$-y = -6x + 6$$

2) $-6x - 2y = -2$

$$2x - 3y = 8$$

3) $5x - 4y = 8$

$$-6x + y = -21$$

4) $10x - 4y = -24$

$$-x - 20y = -18$$

5) $25x + 3y = -13$

$$12x - 6y = -36$$

6) $x - 8y = -7$

$$6x + 4y = 10$$

7) $-6x + 16y = 4$

$$5x + y = 11$$

8) $2x + 3y = 10$

$$4x + 6y = -20$$

Systems of Equations Word Problems

✎ Solve.

1) A school of 210 students went on a field trip. They took 15 vehicles, some vans and some minibuses. Find the number of vans and the number of minibuses they took if each van holds 8 students and each minibus hold 18 students.

2) The difference of two numbers is 14. Their sum is 50. Find the numbers.

3) A farmhouse shelters 15 animals, some are pigs, and some are gooses. Altogether there are 48 legs. How many of each animal are there?

4) The sum of the digits of a certain two–digit number is 9. Reversing it's increasing the number by 9. What is the number?

5) The difference of two numbers is 5. Their sum is 19. Find the numbers.

Graphing Lines of Equations

✎ Sketch the graph of each line

1) $y = 3x - 2$

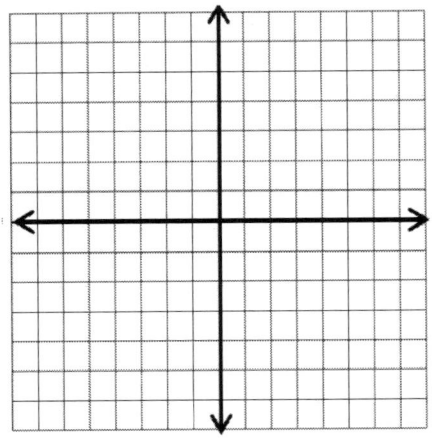

2) $y = -\dfrac{1}{4}x + \dfrac{2}{5}$

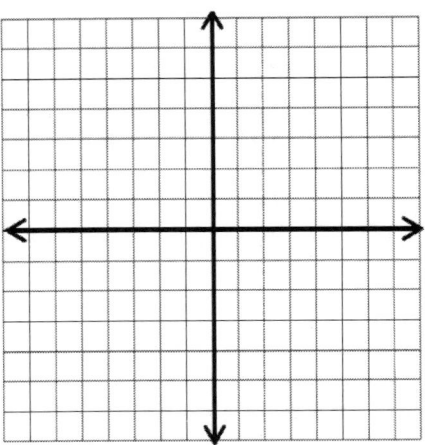

3) $4x - 2y = 6$

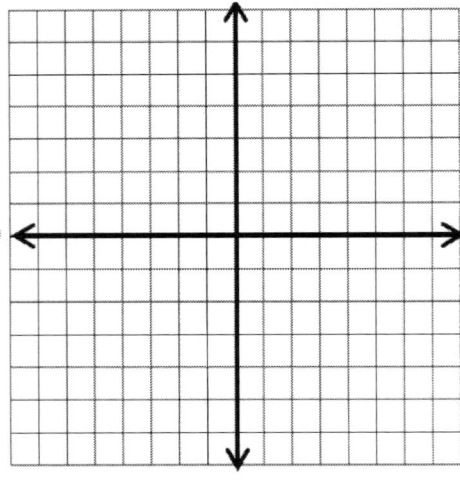

4) $-x - y = 3$

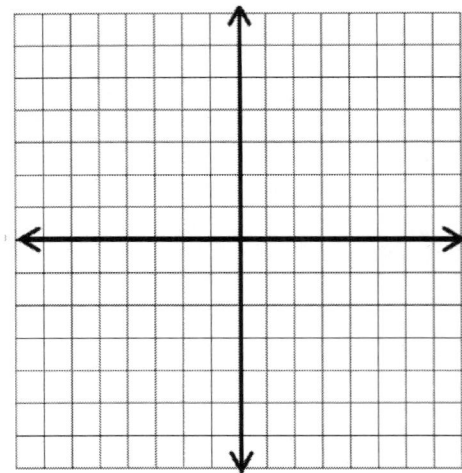

Graphing Linear Inequalities

✎ Sketch the graph of each linear inequality.

1) $y \leq 2x - \frac{2}{3}$

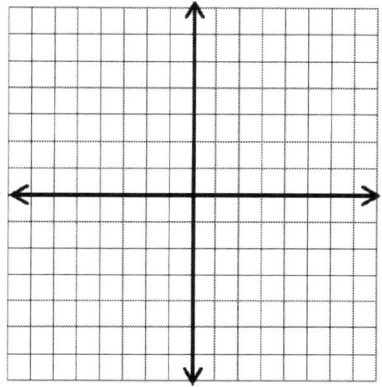

2) $-2x - 3y > \frac{1}{2}$

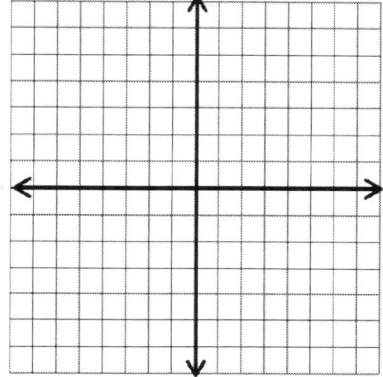

4) $-\frac{1}{2}x + 3y \geq -5$

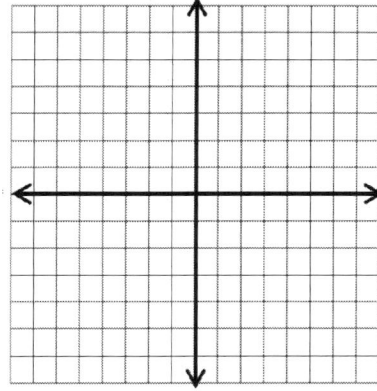

5) $3x - \frac{2}{7}y < 4$

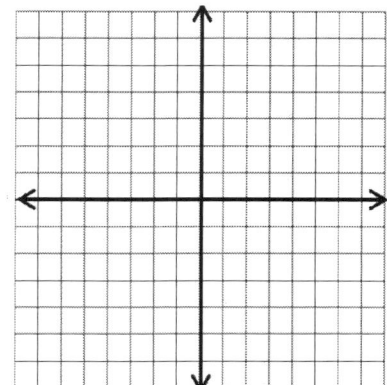

Finding Distance of Two Points

✍ Find the midpoint of the line segment with the given endpoints.

1) $(1, -3), (2, -2)$

2) $(0, 2), (-1, -4)$

3) $(2, 5), (3, -5)$

4) $(1.5, 2.5), (3, 5)$

5) $(-2, 0), (0, -2)$

6) $(5, -4), (2, 5)$

7) $(1, 2), (-1, -2)$

8) $(7, 0.5), (-2, 0)$

9) $(-4, 6), (-3, -1)$

10) $(3.1, -2.7), (-1.2, 3)$

11) $(5.2, 6.3), (4, -2)$

12) $(4, 6), (-2, 4)$

✍ Find the distance between each pair of points.

1) $(4, -1), (2, -1)$

2) $(5, -2), (3, 4)$

3) $(-3, -7), (-1, 2)$

4) $(7, -3), (-5, 0)$

5) $(-8, -5), (-3, -5)$

6) $(11, 1.5), (-2, -8)$

7) $(0, 6), (1, 3)$

8) $(9, 8), (-4, -3)$

9) $(2, 3), (-2, -3)$

10) $(-4, -2), (4, 2)$

11) $(-12, -15), (-7, -9)$

12) $(3, -4), (0, 0)$

Answers of Worksheets – Chapter 7

One–Step Equations

1) 10

2) 48

3) − 10

4) 9

5) − 10

6) − 7

7) − 11

8) -6

9) 5

10) − 6

11) − 2

12) − 15

13) 5

14) -4

15) -11

16) 40

17) 5

18) − 4

19) 34

20) 20

Two–Step Equations

1) 1

2) -4

3) 2

4) -4

5) 1

6) $\frac{5}{3}$

7) $-\frac{81}{6}$

8) 0.9

9) 10

10) 7

11) − 147

12) − 20

13) 50

14) -144

15) − 28

16) 22

17) 4

18) − 31

19) − 61

20) − 16

Multi–Step Equations

1) 5

2) 2

3) 2

4) 15

5) − 2

6) 4

7) 21

8) $\frac{4}{3}$

9) 7

10) − 8

11) − 6

12) 7

13) -20

14) 4

15) 3

16) -4

17) 1 8

18) − 1

19) − 2

20) 0

Graphing Single–Variable Inequalities

1) $4 \geq x$

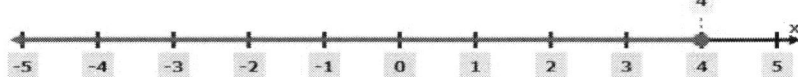

2) $x < -2$

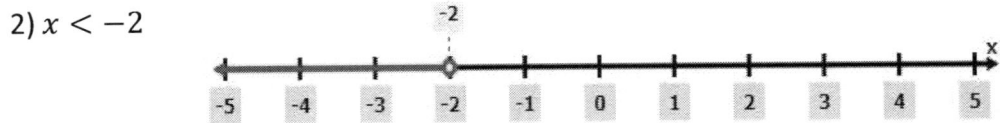

3) $-3 < x$

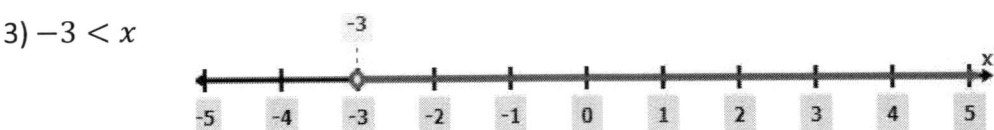

4) $-x \geq 1$

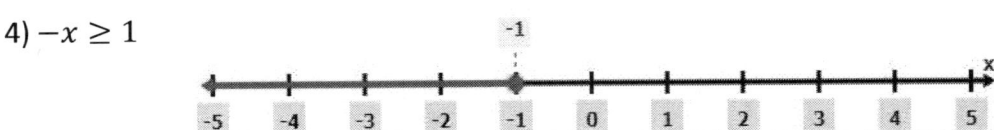

5) $x > 2$

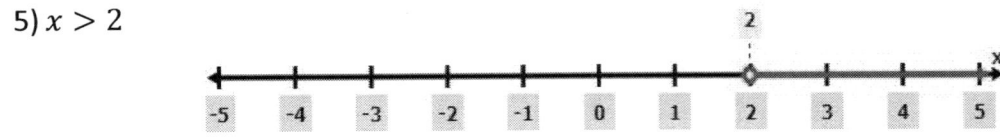

6) $-0.5 < x$

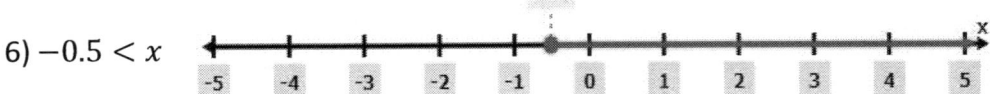

One–Step Inequalities

1)

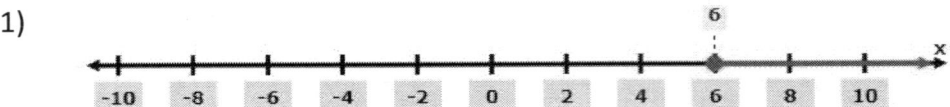

2)

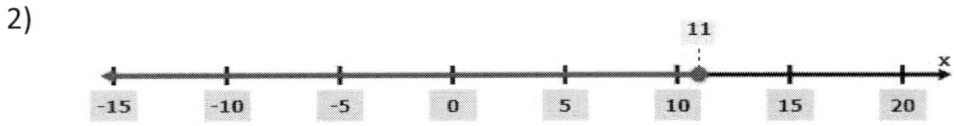

3)

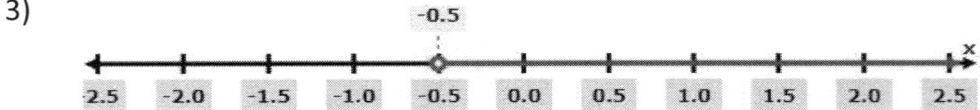

4)

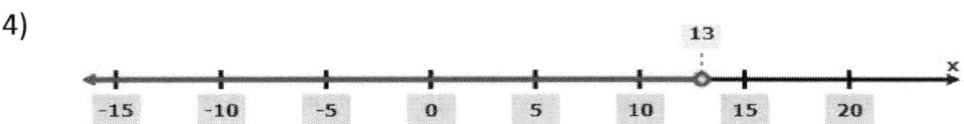

5)

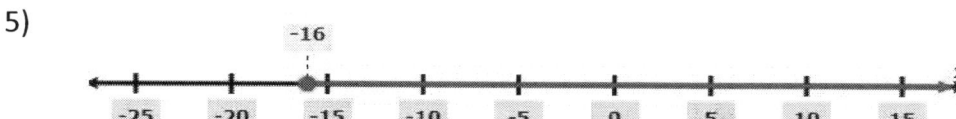

6)

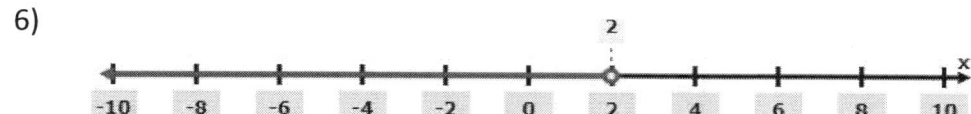

7)

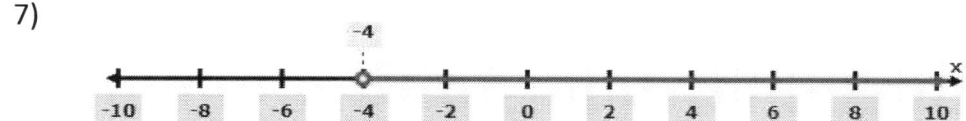

Two–Step inequalities.

1) $x \leq 3$

2) $x \leq 4$

3) $x \leq 3$

4) $x \geq 2$

5) $x \geq 4$

6) $x \leq 5$

7) $x \leq 1$

8) $x \leq 1$

9) $x \leq 1$

10) $x \leq 1$

11) $x < 9.5$

12) $x < 12$

13) $x \geq 3$

14) $x < 3$

15) $x \geq 8$

16) $x < 4$

Multi–Step inequalities.

1) $x < 7$

2) $x \leq 4$

3) $x > 5$

4) $x < 0$

5) $x < 9$

6) $x \leq 17$

Solving Systems of Equations by Substitution

1) (0, 1)

2) (0, −3)

3) (1, −6)

4) (1, 2)

5) (3, 2)

6) (-2, 3)

7) (-4, 1)

8) $(5, \frac{3}{2})$

Solving Systems of Equations by Elimination

1) (−1, 0)

2) (1, −2)

3) (4, 3)

4) (-2, 1)

5) (−1, 4)

6) (1, 1)

7) (2, 1)

8) No solution

Systems of Equations Word Problems

1) There are 6 van and 9 minibuses.

2) 32 and 18

3) There are 9 pigs and 6 gooses.

4) 45

5) 12 and 7.

Writing Linear Equations

1) $y = -\frac{1}{2}x + 5.5$

2) $y = 3x + 7$

3) $y = \frac{1}{4}x + \frac{7}{8}$

4) $y = -\frac{1}{6}x - \frac{23}{8}$

5) $y = -7x - 7$

6) $y = -3x + 15$

7) $y = \frac{1}{2}x - \frac{1}{2}$

8) $y = -\frac{1}{3}x + \frac{4}{5}$

9) $y = 3x - 15$

10) $y = 5$

11) $y = 2x$

12) $y = x - 2$

Graphing Lines Using Slope–Intercept Form

1) f(x)=3x−2   2) f(x)=−1/4 x+2/5

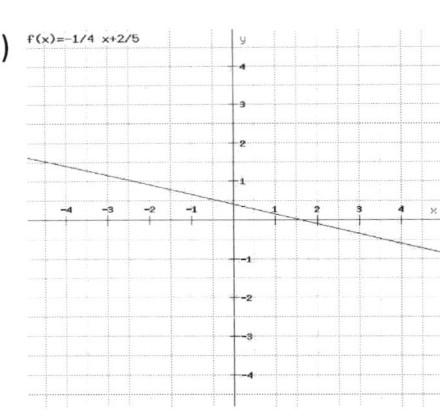

3)

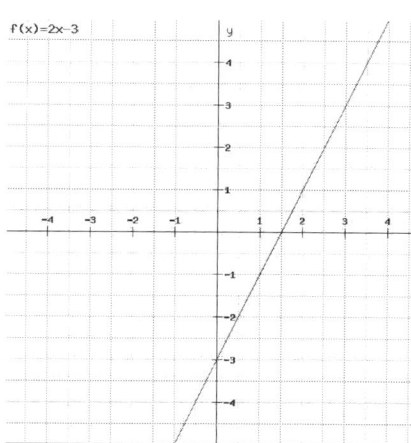

4)

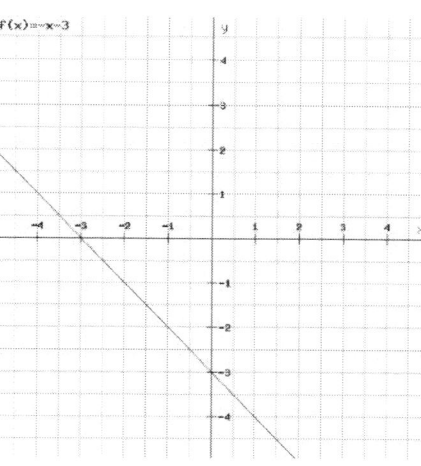

Graphing Linear Inequalities

1)

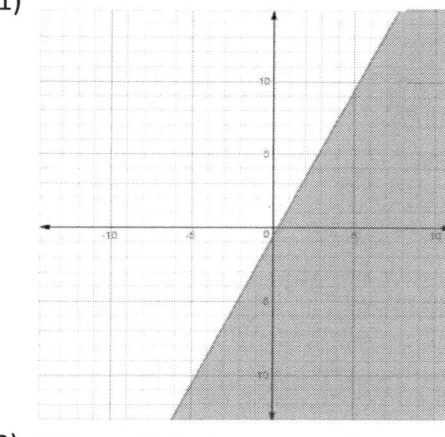

2)

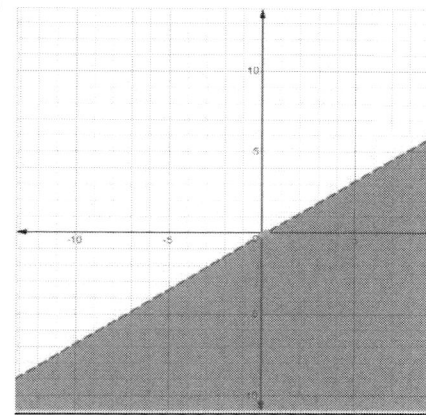

3)

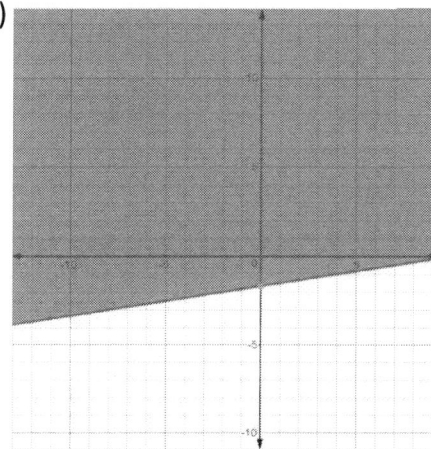

4)

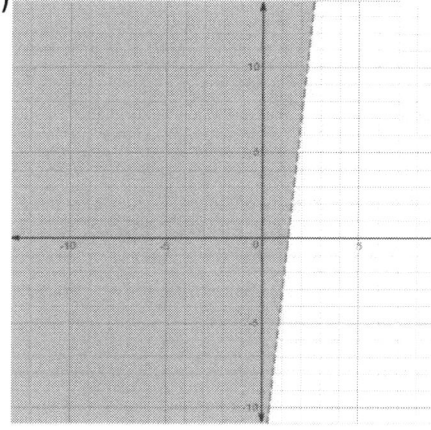

Finding Midpoint

1) $(1.5, -2.5)$

2) $(-1, -1)$

3) $(2.5, 0)$

4) $(2.25, 3.75)$

5) $(-2, -2)$

6) $(3.5, 0.5)$

7) $(0, 0)$

8) $(2.5, 0.25)$

9) $(-3.5, 2.5)$

10) $(0.95, 0.15)$

11) $(4.60, 2.15)$

12) $(1, 5)$

Finding Distance of Two Points

1) 2

2) 6.32

3) 9.22

4) 12.37

5) 5

6) 16.1

7) 3.16

8) 17.03

9) 7.21

10) 8.94

11) 7.81

12) 5

Chapter 8: Polynomials

Topics that you'll learn in this chapter:

- ✓ Classifying Polynomials

- ✓ Writing Polynomials in Standard Form

- ✓ Simplifying Polynomials

- ✓ Adding and Subtracting Polynomials

- ✓ Multiplying and Dividing Monomials

- ✓ Multiplying a Polynomial and a Monomial

- ✓ Multiplying Binomials

- ✓ Factoring Trinomials

- ✓ Operations with Polynomials

Mathematics – the unshaken Foundation of Sciences, and the plentiful Fountain of Advantage to human affairs. — *Isaac Barrow*

Unlock problems

❖ Classifying Polynomials

Name	Degree	Example
constant	0	4
linear	1	$2x$
quadratic	2	$x^2 + 5x + 6$
cubic	3	$x^3 - x^2 + 4x + 8$
quartic	4	$x^4 + 3x^3 - x^2 + 2x + 6$
quantic	5	$x^5 - 2x^4 + x^3 - x^2 + x + 10$

❖ Writing Polynomials in Standard Form

✓ A polynomial function $f(x)$ of degree n is of the form

$$f(x) = a_n x^n + a_{n-1} x^{n-1} + \ldots + a_2 x^2 + a_1 x + a_0$$

The first term is the one with the biggest power!

Example:

$$2x^2 - 4x^3 - x = -4x^3 + 2x^2 - x$$

❖ Simplifying Polynomials

✓ Find "like" terms. (they have same variables with same power).

✓ Add or Subtract "like" terms using PEMDAS operation.

Example:

$$2x^5 - 3x^3 + 8x^2 - 2x^5 = -3x^3 + 8x^2$$

❖ Adding and Subtracting Polynomials

✓ Adding polynomials is just a matter of combining like terms, with some order of operations considerations thrown in.

✓ Be careful with the minus signs, and don't confuse addition and multiplication!

Example:

$$(3x^3 - 1) - (4x^3 + 2) = -x^3 - 3$$

❖ Multiplying Monomials

✓ A monomial is a polynomial with just one term, like $2x$ or $7y$.

✓ When you Multiply two monomials you need to Multiply their coefficients and then Multiply their variables.

✓ In case of exponents with the same base, you need to Add their powers.

Example:

$$2u^3 \times (-3u) = -6u^4$$

❖ Dividing Monomials

✓ When you divide two monomials you need to divide their coefficients and then divide their variables.

✓ In case of exponents with the same base, you need to subtract their powers.

Example: $\dfrac{16a^{-2}}{4a^2} = \dfrac{4}{a^4}$

❖ Multiplying a Polynomial and a Monomial

- ✓ When multiplying monomials, use the product rule for exponents.
- ✓ When multiplying a monomial by a polynomial, use the distributive property.

  $a \times (b + c) = a \times b + a \times c$

Example:

$2x\,(8x - 2) = 16x^2 - 4x$

❖ Multiplying Binomials

- ✓ Use "FOIL". (First–Out–In–Last)

  $(x + a)(x + b) = x^2 + (a + b)x + ab$

Example:

$(x + 2)(x - 3) = x^2 - x - 6$

❖ Factoring Trinomials

- ✓ "FOIL"

  $(x + a)(x + b) = x^2 + (b + a)x + ab$

- ✓ "Difference of Squares"

  $a^2 - b^2 = (a + b)(a - b)$

  $a^2 + 2ab + b^2 = (a + b)(a + b)$

  $a^2 - 2ab + b^2 = (a - b)(a - b)$

- ✓ "Reverse FOIL"

  $x^2 + (a + b)x + ab = (x + a)(x + b)$

Example:

$$x^2 + 5x + 6 = (x + 2)(x + 3)$$

❖ Operations with Polynomials

✓ When multiplying a monomial by a polynomial, use the distributive property.

$$a \times (b + c) = a \times b + a \times c$$

Example: $5(6x - 1) = 30x - 5$

Classifying Polynomials

✍ **Name each polynomial by degree and number of terms.**

1) $x + 4$

2) -8

3) $-5x^4$

4) $9x^2 - 8x^3$

5) $2x - 1$

6) $8x^5$

7) $9x^2 - x$

8) $-5x^4 + 4x^3 - x^2 - 6x$

9) $x - 6x^2 + 4x^3 - 5x^4$

10) $4x^6 + 5x^5 - x^4$

11) $-4 + 2x^2 + x$

12) $9x^6 - 8$

13) $7x^5 + 10x^4 + 3x + 9x^7$

14) $4x^6 + 3x^2 - 5x^4$

✍ **Write each polynomial in standard form**

1) $x^2 - 4x^3$

2) $x^2 + x - x3$

3) $12 - 7x + 9x^4$

4) $x^2 + 12x - 8x3$

5) $x(x + 2) - (x + 2)$

6) $x^2 + x + 13 - 8x^2 - 4x$

7) $10x^5 + x^3 - 3x^5 - 4x^3$

8) $x(x + 6 - 8x^2)$

9) $x(x^5 + 2x^3)$

10) $(x + 4)(x + 2)$

11) $(x + 3)^2$

12) $(x - 5)(2x + 3)$

13) $x(1 + 3x^2 + 2x)$

14) $x(2 - x + 2x^3)$

Simplifying Polynomials

✎ **Simplify each expression.**

1) $5 - 3x^2 + 7x^2 - 2x^3 + 6$

2) $23x^5 - 3x^5 + 7x^2 - 23x^5$

3) $(-2)(x^6 + 9) - 6(10 + x^6)$

4) $3(2x^2 + 4x^2 + 3x^3) - 9x^3 + 17$

5) $3 - 6x^2 + 8x^2 - 13x^3 + 26$

6) $x^2 - 2x + 3x^3 + 16x - 20x$

7) $(x - 6)(3x - 5)$

8) $(12x + y)^2$

9) $(12x^3 + 28x^2 + 10x + 4) \div (x + 2)$

10) $(2x + 12x^2 - 2) \div (2x + 1)$

11) $(x^3 - 1) + (4x^3 - 3x^3)$

12) $(x - 2)(x + 3)$

13) $(2x + 6)(2x - 6)$

14) $(x^2 - 3x) + (5x - 5 - 8x^2)$

Adding and Subtracting Polynomials

1) $(4x^3 + 8) - (9 + 5x^3)$

2) $(x^3 + 8) + (x^3 - 8)$

3) $(2x^2 + 5x^3) - (5x^3 + 9)$

4) $(5x^2 - 2x) + (3x - 6x^2)$

5) $(12x - 7x^3) - (4x^3 + 4x)$

6) $(2x^3 - x^2) - (3x^2 - 4x^3)$

7) $(x^2 - 8) + (8x^2 - 3x^3)$

8) $(x^3 + x^4) - (x^4 + 5x^3)$

9) $(-10x^4 + 12x^5 + x^3) + (14x^3 + 10x^5 + 16x^4)$

10) $(12x^2 - 6x^5 - 2x) + (-10x^2 + 11x^5 - 9x)$

11) $(35 + 8x^5 - 4x^2) + (7x^4 + 2x^5) - (27 - 4x^4)$

12) $(4x^5 - 3x^3 - 3x) + (3x + 10x^4 - 12) + (2x^2 + x^3 + 10)$

Multiplying Monomials

1) $4xy^2z \times 3z^2$

2) $6xy \times 2x^2y$

3) $6pq^3 \times (-3p^4q)$

4) $6s^4t^2 \times st^5$

5) $12p^3 \times (-4p^4)$

6) $-4p^2q^3r \times 4pq^2r^3$

7) $(-4)(-24a^6b)$

8) $2u^4v^2 \times (-10u^2v^3)$

9) $5u^3 \times (-3u)$

10) $-8xy^2 \times 4x^2y$

11) $24y^2z^3 \times (-y^2z)$

12) $10a^2bc^2 \times 3abc^2$

Divide Monomials

1) $(8x^4y^6) \div (4x^3y^4)$

2) $\dfrac{60x^5y^8}{40x^7y^{11}}$

3) $(14x^4) \div (4x^9)$

4) $\dfrac{60x^{12}y^9}{20x^6y^7}$

5) $\dfrac{1}{(-2x^{-4}y^2)^5}$

6) $\dfrac{85x^{18}y^7}{5x^9y^2}$

7) $(12x^2y^9) \div (6x^9y^{12})$

8) $\dfrac{200x^3y^8}{20x^3y^7}$

9) $\dfrac{(3x^{-5}y^4z)^{-2}}{9x^6y^{-1}}$

10) $\dfrac{-18x^{17}y^{13}}{3x^6y^9}$

11) $\dfrac{-5x^{-3}y^{-1}}{-4x^{-4}y^3}$

12) $\dfrac{-81x^8y^{10}}{9x^3y^7}$

Multiply a Polynomial and a Monomial

1) $4(3x - 4y)$

2) $9x(3x + 5y)$

3) $8x(8x - 5)$

4) $12x(3x + 9)$

5) $12x(2x - 2y)$

6) $3x(5x - 6y)$

7) $x(2x^2 - 3x + 8)$

8) $12x(2x + 4y)$

9) $30(2x^2 - 8x - 5)$

10) $6x^3(3x^2 - x + 1)$

11) $8x^2(4x^2 - 5xy + y^2)$

12) $x^2(3x^2 - 5x + 10)$

13) $2x^3(x^2 + 6x - 2)$

14) $4x(3x^2 - 4xy + 2y^2)$

Multiply Binomials

1) $(x - 2)(5x + 2)$

2) $(3x - 2)(x + 5)$

3) $(x + 2)(x + 8)$

4) $(x^2 + 3)(x^2 - 3)$

5) $(x - 3)(x + 6)$

6) $(x - 6)(2x + 6)$

7) $(x - 4)(3x - 3)$

8) $(x - 5)(x - 4)$

9) $(x + 5)(2x + 5)$

10) $(x - 6)(3x + 6)$

11) $(x - 8)(x + 8)$

12) $(x - 4)(4x + 8)$

13) $(2x - 6)(2x + 6)$

14) $(x + 7)(x - 2)$

15) $(x - 7)(x + 7)$

16) $(4x + 4)(4x - 3)$

Factor Trinomials

1) $x^2 - 6x + 8$

2) $x^2 - 5x - 14$

3) $x^2 - 10x - 24$

4) $2x^2 + 3x - 9$

5) $x^2 - 16x + 48$

6) $x^2 + 3x - 18$

7) $3x^2 + 7x + 2$

8) $x^2 - 3x - 10$

9) $8x^2 + 22x - 6$

10) $x^2 + 22x + 121$

11) $64x^2 + 16xy + 4y^2$

12) $6x^2 - 20x + 16$

13) $x^2 - 12x + 36$

14) $25x^2 + 20x + 4$

Operations with Polynomials

1) $2x^2(4x - 3)$

2) $4x^2(6x - 3)$

3) $-5(5x - 3)$

4) $4x^3(-4x + 6)$

5) $7(7x + 2)$

6) $9(3x + 7)$

7) $4(7x + 1)$

8) $-6x^4(x - 4)$

9) $8(x^2 - 2x + 3)$

10) $2(4x^2 - 2x + 1)$

11) $2(4x^2 + 3x - 2)$

12) $7x(2x^2 + 3x + 8)$

13) $(8x + 1)(2x - 1)$

14) $(x + 5)(3x - 5)$

15) $(6x - 4)(3x - 6)$

16) $(x - 4)(3x + 6)$

Answers of Worksheets – Chapter 8

Classifying Polynomials

1) Linear monomial

2) Constant monomial

3) Quantic monomial

4) cubic binomial

5) linear binomial

6) Quantic monomial

7) Quadratic binomial

8) Quartic polynomial with four terms

9) Quartic polynomial with four terms

10) Sixth degree trinomial

11) Quadratic trinomial

12) Sixth degree binomial

13) Seventh degree polynomial with four terms

14) Sixth degree trinomial

Writing Polynomials in Standard Form

1) $-4x^3 + x^2$

2) $-x^3 + x^2 + x$

3) $9x^4 - 7x + 12$

4) $-8x^3 + x^2 + 12x$

5) $x^2 + x - 2$

6) $-7x^2 - 3x + 13$

7) $7x^5 - 3x^3$

8) $-8x^3 + x^2 + 6x$

9) $x^6 + 2x^4$

10) $x^2 + 6x + 8$

11) $x^2 + 6x + 9$

12) $2x^2 - 7x - 15$

13) $3x^3 + 2x^2 + x$

14) $2x^4 - x^2 + 2x$

Simplifying Polynomials

1) $-2x^3 + 4x^2 + 11$

2) $-3x^3 + 7x^2$

3) $-8x^6 - 78$

4) $18x^2 + 17$

5) $-13x^3 + 2x^2 + 29$

6) $3x^3 + x^2 - 6x$

7) $3x^2 - 23x + 30$

8) $144x^2 + 24xy + y^2$

9) $12x^2 + 4x + 2$

10) $6x - 2$

11) $2x^3 - 1$

12) $x^2 + x - 6$

13) $4x^2 - 36$

14) $-7x^2 + 2x - 5$

Adding and Subtracting Polynomials

1) $-x^3 - 1$

2) $2x^3$

3) $2x^2 - 9$

4) $-x^2 + x$

5) $-11x^3 + 8x$

6) $6x^3 - 4x^2$

7) $-3x^3 + 9x^2 - 8$

8) $-4x^3$

9) $22x^5 + 6x^4 + 15x^3$

10) $5x^5 + 2x^2 - 11x$

11) $10x^5 + 11x^4 - 4x^2 + 8$

12) $4x^5 + 10x^4 - 2x^3 + 2x^2 - 2$

Multiply Monomials

1) $12xy^2z^3$

2) $12x^3y^2$

3) $-18p^5q^4$

4) $6s^5t^7$

5) $-48p^7$

6) $-16p^3q^5r^4$

7) $96a^6b$

8) $-20u^6v^5$

9) $-15u^4$

10) $-32x^3y^3$

11) $-24y^4z^4$

12) $30a^3b^2c^4$

Divide Monomials

1) $2xy^7$

2) $\dfrac{1.5}{x^2y^3}$

3) $\dfrac{3.5}{x^5}$

4) $3x^6y^2$

5) $-\dfrac{x^{20}}{32y^{10}}$

6) $17x^9y^5$

7) $\dfrac{2}{x^7y^3}$

8) $10y$

9) $\dfrac{x^4}{81y^7z^2}$

10) $-6x^{11}y^4$

11) $\dfrac{1.25x}{y^4}$

12) $-9x^5y^3$

Multiply a Polynomial and a Monomial

1) $12x - 16y$

2) $72x^2 + 45xy$

3) $64x^2 - 40x$

4) $36x^2 + 108x$

5) $24x^2 - 24xy$

6) $15x^2 - 18xy$

7) $2x^3 - 3x^2 + 8x$

8) $24x^2 + 48xy$

9) $60x^2 - 240x - 150$

10) $18x^5 - 6x^4 + 6x^3$

11) $32x^4 - 40x^3y + 8y^2x^2$

12) $3x^4 - 5x^3 + 10x^2$

13) $2x^5 + 12x^4 - 4x^3$

14) $12x^3 - 16x^2y + 8xy^2$

Multiplying Binomials

1) $5x^2 - 8x - 4$

2) $3x^2 + 13x - 10$

3) $x^2 + 10x + 16$

4) $x^4 - 9$

5) $x^2 + 3x - 18$

6) $2x^2 - 6x - 36$

7) $3x^2 - 15x + 12$

8) $x^2 - 9x + 20$

9) $2x^2 + 15x + 25$

10) $3x^2 - 13x - 36$

11) $x^2 - 64$

12) $3x^2 - 13x - 32$

13) $4x^2 - 36$

14) $x^2 + 5x - 14$

15) $x^2 - 49$

16) $16x^2 + 4x - 12$

Factoring Trinomials

1) $(x - 4)(x - 2)$

2) $(x + 2)(x - 7)$

3) $(x + 2)(x - 12)$

4) $(x + 3)(2x - 3)$

5) $(x - 14)(x - 2)$

6) $(x - 3)(x + 6)$

7) $(3x + 1)(x + 2)$

8) $(x - 5)(x + 2)$

9) $(4x - 1)(2x + 6)$

10) $(x + 11)(x + 11)$

11) $(8x + 2y)(8x + 2y)$

12) $(2x - 4)(3x - 4)$

13) $(x - 6)(x - 6)$

14) $(5x + 2)(5x + 2)$

Operations with Polynomials

1) $8x^3 - 6x^2$

2) $24x^3 - 12x^2$

3) $-25x + 15$

4) $-16x^4 + 24x^3$

5) $49x + 14$

6) $18x + 63$

7) $28x + 4$

8) $-6x^5 + 24x^4$

9) $8x^2 - 16x + 24$

10) $8x^2 - 4x + 2$

11) $8x^2 + 6x - 4$

12) $14x^3 + 21x^2 + 56x$

13) $16x^2 - 6x - 1$

14) $3x^2 + 10x - 25$

15) $18x^2 - 48x + 24$

16) $3x^2 - 6x - 24$

Chapter 9: Functions

Topics that you'll learn in this chapter:

- ✓ Relations and Functions
- ✓ Adding and Subtracting Functions
- ✓ Multiplying and Dividing Functions
- ✓ Finding Slope and Rate of Change
- ✓ Find the X–intercept and Y–intercept
- ✓ Graphing Lines Using Slope–Intercept Form
- ✓ Graphing Lines Using Standard Form
- ✓ Writing Linear Equations
- ✓ Write an Equation from a Graph
- ✓ Equations of horizontal and vertical lines
- ✓ Equation of parallel or perpendicular lines
- ✓ Solving Quadratic Functions

Without mathematics, there's nothing you can do. Everything around you are mathematics. Everything around you are numbers." – Shakuntala Devi

Unlock Problems

❖ Relation and Functions

  ✓ **RELATION:** A relationship between two sets of elements (input and output) in some way, and input can have as many as outputs.

  ✓ **FUNCTION:** A relationship between two sets of elements (input and output) and only and exactly one output related to one input.

  A function is a type of relation. relation may not be a function.

The Vertical Line Test

  A graphical method that we can observe the cross between the graph and vertical line.

  The graph is a function if and only if the intersection point is one.

Example:

  The graph is not a function

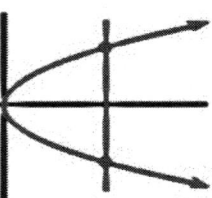

❖ Slope

  ✓ Slope of a line:

  $$\frac{y_2 - y_1}{x_2 - x_1} = \frac{rise}{run}$$

Example:

  $(2, -10), (3, 6)$

$$\text{slope} = 16$$

❖ Rate of Change

- ✓ Slope can be described as "rate of change".
- ✓ Rate of change is a ratio between a change in one variable comparing to a corresponding change in another variable.

Example:

❖ x and y–intercept

- ✓ x-intercept: is a point in the equation where the value of y is zero.
- ✓ y-intercept: is a point in the equation where the value of x is zero.

❖ Slope–intercept Form

- ✓ Using the slope m and the
- ✓ y-intercept b, then the equation of the line is:

$$y = mx + b$$

Example:

$$y = -10 + 2x$$
$$m = 2$$

❖ Point–slope Form

- ✓ Using the slope m and a point $(x1, y1)$ on the line, the equation of the line is

$$(y - y1) = m(x - x1)$$

Example:

$$y = 2(x + 3)$$
$$m = 2, (-3, 0)$$

❖ Writing Linear Equations

✓ The equation of a line:

$$y = mx + b$$

✓ 1– Identify the slope.

✓ 2– Find the y–intercept. This can be done by substituting the slope and the coordinates of a point (x, y) on the line.

Example:

through:

$$(-4, -2), (-3, 5)$$
$$y = 7x + 26$$

❖ Write an Equation from a Graph

✓ An equation in the slope-intercept form is written as

$$y = mx + b$$

While m is the slope of the line and b is the y-intercept.

❖ Graphing Lines Using Slope–Intercept Form

✓ given the slope m and the y–intercept b, then the equation of the line is: $y = mx + b$.

Example:

$y = 8x - 3$

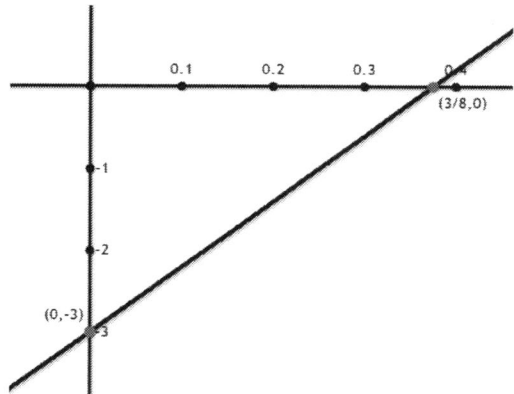

❖ Equation of Parallel or Perpendicular Lines

✓ Parallel lines: are distinct lines with the same slope. For example: if the following lines are parallel:

$y = m_1 x + b_1$

$y = m_2 x + b_2$

Then, $m_1 = m_2$ and $b_1 \neq b_2$.

✓ Perpendicular Lines: A pair of lines is perpendicular if the lines meet at 90° angle.

$y = m_1 x + b_1$

$y = m_2 x + b_2$

✓ the two lines are perpendicular if, $m_1 = -\dfrac{1}{m_2}$, that is, if the slopes are negative reciprocals of each other.

❖ Equations of Horizontal and Vertical Lines

✓ The slope of horizontal lines is 0. Thus, the equation of horizontal lines becomes: $y = b$

✓ The slope of vertical lines is undefined and the equation for a vertical line is: $x = a$

❖ Function Notation

✓ Function notation is the way a function is written. It is meant to be a precise way of giving information about the function without a rather lengthy written explanation. The most popular function notation is $f(x)$ which is read "f of x".

Example: $f(x) = 12x$

❖ Function Notation

✓ Function notation is the way a function is written. It is meant to be a precise way of giving information about the function without a rather lengthy written explanation. The most popular function notation is $f(x)$ which is read "f of x".

Example: $f(x) = 12x$

❖ Multiplying and Dividing Functions

✓ Just like we can multiply and divide numbers, we can multiply and divide functions. For example, if we had functions f and g, we could create two new functions $f \cdot g$, and $\frac{f}{g}$.

Example:

$$f(x) = 2x \; ; \; g(x) = x^2 + x$$
$$(f \cdot g)(x) = f(x) \cdot g(x) = 2x(x^2 + x)$$
$$2x^3 + 2x^2$$

❖ Composition of Functions

✓ The term "composition of functions" (or "composite function") refers to the combining together of two or more functions in a manner where the output from one function becomes the input for the next function.

✓ The notation used for composition is:

$$(f \circ g)(x) = f(g(x))$$

Example:

Using $f(x) = x + 1$ and $g(x) = 2x$, find: $(f \circ g)(1)$

$(f \circ g)(x) = 2x + 1$

$(f \circ g)(1) = 3$

Relation and Functions

✍State the domain and range of each relation. Then determine whether each relation is a function.

1)

Function:

....

Domain:

....

Range:

....

2)

Function:

....

Domain:

....

Range:

....

x	y
2	2
0	0
−1	−1
4	−1
5	1

3)

Function:

....

Domain:

....

Range: ....

4) $\{(2, -1), (3, -2), (0, 1), (3, 0), (1, 1)\}$

Function: ....

Domain:

....

Range:

....

5)

Function:

....

Domain:

....

Range:

....

6)

Function:

....

Domain:

....

Range:

....

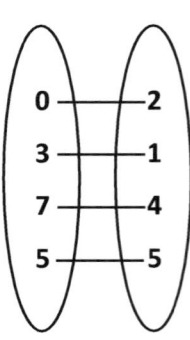

Slope and Rate of Change

✍ **Find the slope of the line through each pair of points.**

1) $(4, 1), (2, 5)$

2) $(1, -6), (-5, 3)$

3) $(2, -7), (5, -8)$

4) $(12, 9), (18, 14)$

5) $(0, -3), (7, -2)$

6) $(11, -7), (13, -5)$

7) $(-3, -5), (-11, -1)$

8) $(0, 0), (6, -1)$

9) $(16, -9), (-4, 4)$

10) $(-8, 6), (-8, 2)$

11) $(-12, -7), (-4, -13)$

12) $(-14, 0), (0, -14)$

✍ **Write the slope–intercept form of the equation of the line through the given points.**

1) Through: $(5, 3), (7, 2)$

2) Through: $(-4, -5), (-3, -2)$

3) Through: $(0.4, 1), (2, 1.4)$

4) Through: $(7, -3), (2.5, 1)$

5) Through: $(-1, 0), (-2, 7)$

6) Through: $(7, -6), (2, 9)$

7) Through: $(9, 4), (7, 3)$

8) Through: $(-0.5, 1), (5.5, -1)$

9) Through: $(4, -3), (8, 9)$

10) Through: $(1, 5), (-2, 5)$

11) Through: $(2, 4), (-1, -2)$

12) Through: $(8, 6), (0, -2)$

Find the value of b*: The line that passes through each pair of points has the given slope.*

1) $(3, -2), (1, b), m = 1$

3) $(-3, b), (3, 7), m = \frac{1}{3}$

2) $(b, -6), (-3, 8), m = -1\frac{2}{5}$

4) $(0, 2), (b, 5), m = -\frac{1}{3}$

Write the slope intercept form of the equation of each line

1)

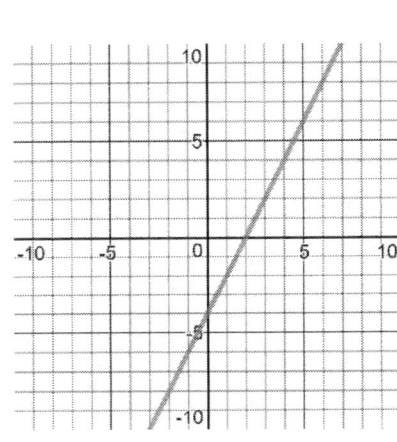

2)

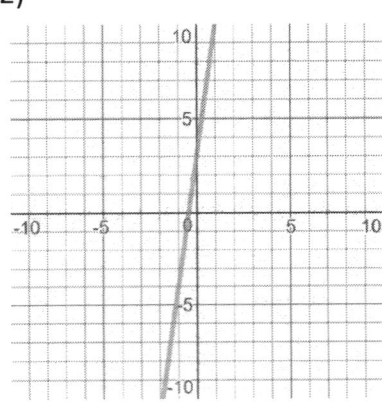

3)

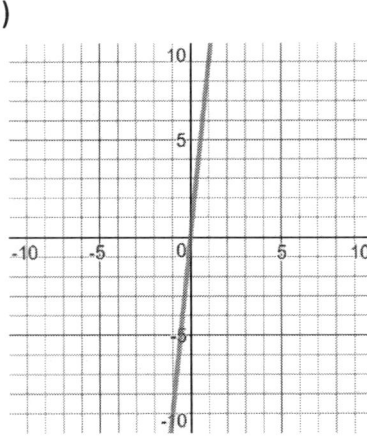

4)

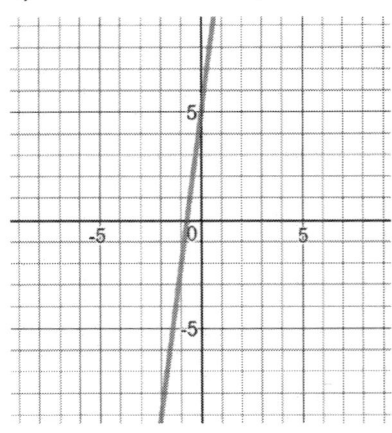

Rate of change

✍ **What is the average rate of change of the function?**

1) $f(x) = 2x^2 + 3$, from $x = 3$ to $x = 7$?

2) $f(x) = -x^2 - 2$, from $x = 0$ to $x = 2$?

3) $f(x) = x^3 + 1$, from $x = 1$ to $x = 3$?

*x*and*y* intercepts

✍**Find the x and y intercepts for the following equations.**

1) $3x + 2y = 12$ 2) $y = x + 7$ 3) $3x = y + 15$

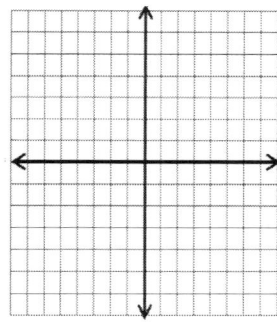   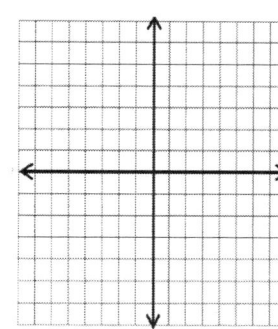   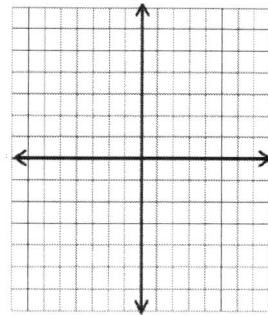

4) $x + y = 0$ 5) $7x - 3y = 5$ 6) $5y - 4x + 8 = 0$

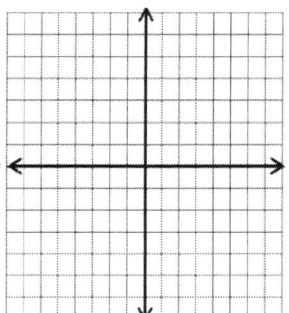   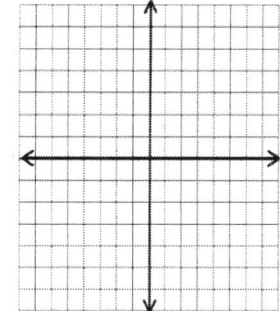   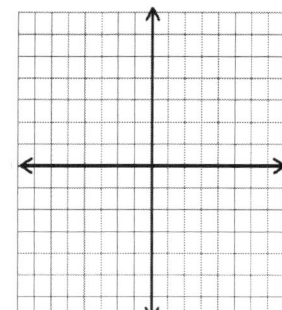

Slope–intercept Form

✎ **Write the slope–intercept form of the equation of each line.**

1) $-12x + y = 5$

2) $-3(5x + y) = 45$

3) $-7x - 14y = -42$

4) $6x + 25 = -4y$

5) $x - 2y = 8$

6) $21x - 15y = -9$

7) $32x - 16y = -64$

8) $9x - 6y + 27 = 0$

9) $-\frac{2}{7}y = -3x + 1$

10) $7 - y - 3x = 0$

11) $-y = -3x - 8$

12) $8x + 4y = -16$

13) $2(x + y + 5) = 0$

14) $y - 5 = x + 6$

15) $4(y + 3) = 5(x - 3)$

16) $\frac{2}{3}y + \frac{1}{3}x + \frac{4}{3} =$

Point–slope Form

✍ **Find the slope of the following lines. Name a point on each line.**

1) $y = 3(x + 2)$

2) $y + 5 = \dfrac{2}{3}(x - 2)$

3) $y + 1 = -2.5x$

4) $y - 5 = \dfrac{1}{2}(x - 4)$

5) $y + 2 = 1.4\,(x + 2)$

6) $y - 7 = -4x$

7) $y - 10 = -4\,(x - 6)$

8) $y + 15 = 0$

9) $y + 8 = 3\,(x + 1)$

10) $y - 11 = -7\,(x - 3)$

✍ **Write an equation in point–slope form for the line that passes through the given point with the slope provided.**

11) $(2, -1), m = 5$

12) $(-2, 4), m = \dfrac{1}{2}$

13) $(0, -8), m = -1$

14) $(a, b), m = m$

15) $(-5, 3), m = 5$

16) $(2, 0), m = -6$

17) $(-6, 8), m = \dfrac{2}{3}$

18) $(-1, 12), m = 0$

19) $\left(-\dfrac{1}{4}, 2\right), m = \dfrac{1}{8}$

20) $(0, 0), m = -4$

Equation of Parallel or Perpendicular Lines

✍ **Write an equation of the line that passes through the given point and is parallel to the given line.**

1) $(-2, -4), 2x + 3y = -6$

2) $(-3, 0), y = x - 3$

3) $(-1, 0), 3y = 7x - 2$

4) $(0, 0), -2y + 6x - 15 = 0$

5) $(2, 13), y + 17 = 0$

6) $(0, 5), -8x - y = -7$

7) $(-3, -2), y = \frac{3}{4}x + 2$

8) $(-1, 3), -6x + 5y = -17$

9) $(5, -3), y = -\frac{3}{5}x - 2$

10) $(-4, -4), 9x + 12y = -24$

✍ **Write an equation of the line that passes through the given point and is perpendicular to the given line.**

11) $(-2, -5), 2x + 3y = -9$

12) $(-\frac{1}{2}, \frac{3}{4}), 3x - 9y = -24$

13) $(3, -7), y = -7$

14) $(8, -4), x = 8$

15) $(0, -4), y = \frac{1}{3}x + 5$

16) $(\frac{2}{5}, \frac{4}{5}), y = -4x - 21$

17) $(-6, 0), y = \frac{3}{2}x - 13$

18) $(1, -3), y = x + 15$

19) $(-2, -2), y = \frac{5}{4}x - 1$

20) $(0, 0), y - 7x + 6 = 0$

Equations of Horizontal and Vertical Lines

✏️ Sketch the graph of each line.

1) $y = 2$

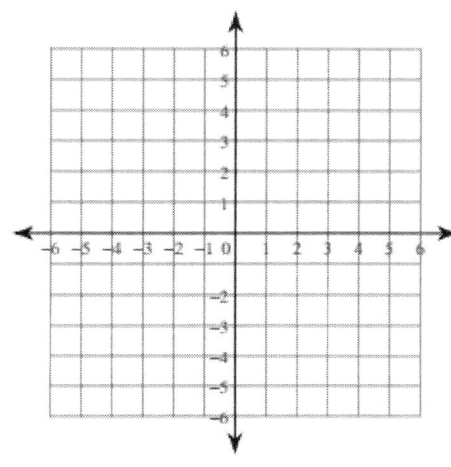

2) $y = 0$

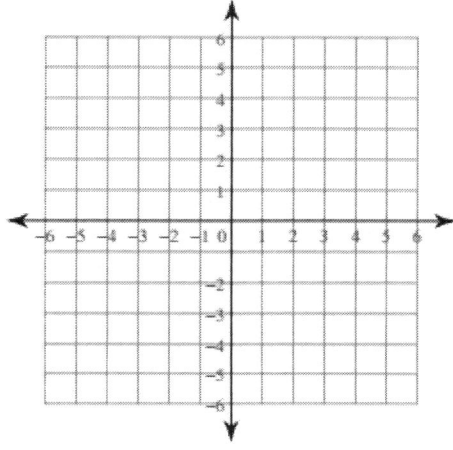

3) $x = 3$

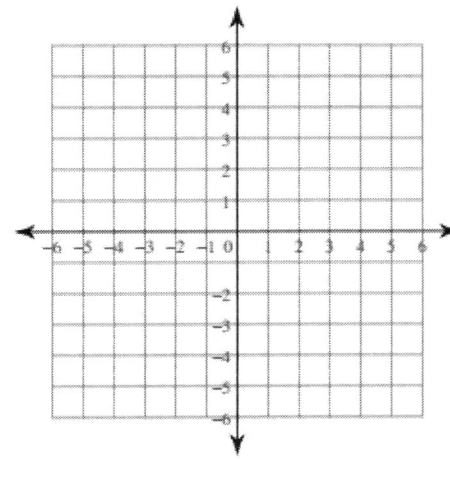

4) $x = -4$

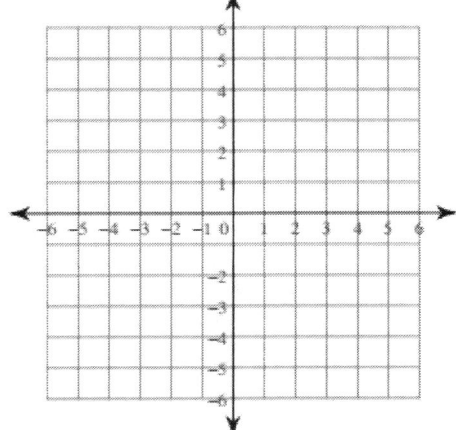

Function Notation

✍Write in function notation.

1) $v = 7t$

2) $r = 2p^2 + p - 1$

3) $h = 14g + 7$

4) $y = 3x - \dfrac{2}{5}$

✍Evaluate each function.

5) $w(x) = 5x + 2$, find $w(3)$

6) $h(n) = n^2 - 4$, find $h(-2)$

7) $h(x) = x^3 - 4$, find $h(-1)$

8) $h(m) = -3m^2 + 4m$, find $h(5)$

9) $f(n) = n^2 - n$, find $g(12)$

10) $g(x) = x^3 + 12x + 8$, find $g(0)$

11) $f(u) = 8u - 3$, find $g(^1/_4)$

12) $h(x) = 2x - 4$, find $h(a)$

13) $h(a) = -6a + 2$, find $h(2b)$

14) $k(a) = -2a + 1$, find $k(a - 3)$

15) $h(x) = x^3 + 2x^2 - 5$, find $h(x^2)$

16) $h(x) = x^2 + 2$, find $h(-\dfrac{x}{2})$

Adding and Subtracting Functions

✎ **Perform the indicated operation.**

1) $h(t) = 3t - 2; g(t) = 3t + 2$

   Find $(h - g)(t)$.

2) $g(a) = -2a^2 + 1; f(a) = a^2 - a + 4$

   Find $(g - f)(a)$.

3) $g(x) = 3x - 4; h(x) = x - 7$

   Find $g(2) - h(2)$.

4) $h(3) = -2x + 1; g(x) = 3x - 5$

   Find $(h + g)(3)$.

5) $f(x) = 4x - 2; g(x) = x^2 + x$

   Find $(f - g)(-1)$.

6) $h(n) = 3n - 3 ; g(n) = n^2 - 2n + 4$

   Find $(h + g)(a)$.

7) $g(x) = -x^2 + 5 - 2x; f(x) = 11 + 3x$

   Find $(g - f)(x)$.

8) $h(x) = x^2 - 8; g(x) = -2x^2 + x$

   Find $(h + g)(t)$.

9) $g(t) = t + 7; f(t) = -2t^2 + t$

   Find $(g - f)(u + 1)$.

10) $k(x) = -2x + 9; h(x) = -x^2 + 2x - 4$

    Find $(k + h)(t - 5)$.

Multiplying and Dividing Functions

✏Perform the indicated operation.

1) $g(a) = 3a - 2; h(a) = 5a - 1$

   Find $(g.h)(-2)$

2) $f(x) = x^3 - 2x^2; g(x) = 3x - 1$

   Find $(f.g)(x)$

3) $g(t) = \frac{1}{2}t^2 + \frac{1}{2}; h(t) = 2t - 6$

   Find $(h.g)(-\frac{1}{2})$

4) $k(n) = n^2 - n; h(n) = 2n^2 + 2$

   Find $(k.h)(1)$

5) $f(a) = 11a - 16; g(a) = 5a + 12$

   Find $(\frac{f}{g})(-2)$

6) $f(x) = x - 1; g(x) = x^2 - 1$

   Find $(\frac{g}{f})(x)$

7) $h(a) = -2a; g(a) = -a^2 - a$

   Find $(\frac{h}{g})(a)$

8) $f(t) = -a + 2; g(t) = a^3 - 1$

   Find $(\frac{2f}{g})(a)$

Composition of Functions

✎Using $f(x) = 3x - 7$, and $g(x) = -x + 1$, find:

1) $f(g(1))$

2) $f(f(0))$

3) $g(f(-5))$

✎Using $f(x) = -2x + 5$, and $g(x) = x - 4$, find:

4) $f(g(-3))$

5) $g(g(2))$

6) $g(f(\frac{1}{2}))$

✎Using $f(x) = 5x - 2a$, and $g(x) = x^2 - 3$, find:

7) $(fog)(-1) = f(g(-1))$

8) $(fof)(4)$

9) $(gof)(-1)$

✎Using $f(x) = -x + 5$, and $g(x) = x + b$, find:

10) $(fog)(x)$

11) $(fog)(x + 2)$

12) $(gof)(x^2)$

Answers of Worksheets – Chapter 9

Relation and Functions

1) No, $D_f = \{2, 4, 6, 8, 10\}$, $R_f = \{4, 8, 12, 16, 20\}$

2) Yes, $D_f = \{2, 0, -1, 4, 5\}$, $R_f = \{2, 0, -1, -1, 1\}$

3) Yes, $D_f = (-\infty, \infty)$, $R_f = \{-2, \infty)$

4) No, $D_f = \{2, 3, 0, 3, 1\}$, $R_f = \{-1, -2, 1, 0, 1\}$

5) No, $D_f = [-2, 2]$, $R_f = [-2, 3]$

6) Yes, $D_f = \{0, 3, 7, 5\}$, $R_f = \{2, 1, 4, 5\}$

Finding Slope

1) -2

2) -1.5

3) $-\frac{1}{3}$

4) $\frac{5}{6}$

5) $\frac{1}{7}$

6) 1

7) $-\frac{1}{2}$

8) $-\frac{1}{6}$

9) $-\frac{13}{20}$

10) Undefined

11) $-\frac{3}{4}$

12) -1

Writing Linear Equations

1) $y = -\frac{1}{2}x + 5.5$

2) $y = 3x + 7$

3) $y = \frac{1}{4}x + \frac{9}{10}$

4) $y = -\frac{8}{9}x + 3\frac{2}{9}$

5) $y = -7x - 7$

6) $y = -3x + 15$

7) $y = \frac{1}{2}x - \frac{1}{2}$

8) $y = -\frac{1}{3}x + \frac{5}{6}$

9) $y = 3x - 15$

10) $y = 5$

11) $y = 2x$

12) $y = x - 2$

Find the value of b

1) -4

2) 7

3) 5

4) -9

Write an equation from a graph

1) $y = 2x - 4$

2) $y = 7x + 3$

3) $y = 9x$

4) $y = 7x + 5$

Rate of change

1) 20 2) -2 3) 13

x–intercept and y–intercept

1) $y - intercept = 6$

$x - intercept = 4$

2) $y - intercept = 7$

$x - intercept = -7$

3) $y - intercept = -15$

$x - intercept = 5$

4) $y - intercept = 0$

$x - intercept = 0$

5) $y - intercept = -\frac{5}{3}$

$x - intercept = \frac{5}{7}$

6) $y - intercept = -\frac{8}{5}$

$x - intercept = 2$

Slope–intercept form

1) $y = 12x + 5$

2) $y = -5x - 15$

3) $y = -\frac{1}{2}x + 3$

4) $y = -\frac{3}{2}x - \frac{25}{4}$

5) $y = \frac{x}{2} - 4$

6) $y = \frac{7}{5}x + \frac{3}{5}$

7) $y = 2x + 4$

8) $y = \frac{3}{2}x + \frac{9}{2}$

9) $y = \frac{21}{2}x - \frac{7}{2}$

10) $y = -3x + 7$

11) $y = 3x + 8$

12) $y = -2x - 4$

13) $y = -x - 5$

14) $y = x + 11$

15) $y = \frac{5}{4}x - \frac{27}{4}$

16) $y = -\frac{1}{2}x - 2$

Point–slope form

1) $m = 3, (-2, 0)$

2) $m = \frac{2}{3}, (5, -3)$

3) 3) $m = -\frac{5}{2}, (0, -1)$

4) $m = \frac{1}{2}, (6, 6)$

5) $m = \frac{14}{10}, (-2, -2)$

6) $m = -4, (0, 7)$

7) $m = -4, (1, 30)$

8) $m = 0, (5, -15)$

9) $m = 3, (0, -5)$

10) $m = -7, (-2, 46)$

11) $y + 1 = 5(x - 2)$

12) $y - 4 = \frac{1}{2}(x + 2)$

13) $y + 8 = -x$

14) $y - b = m(x - a)$

15) $y - 3 = 5(x + 5)$

16) $y = -6(x - 2)$

17) $y - 8 = \frac{2}{3}(x + 6)$

18) $y - 12 = 0$

19) $y - 2 = \frac{1}{8}\left(x + \frac{1}{4}\right)$

20) $y = -4x$

Equation of parallel or perpendicular line.

1) $y = -\frac{2}{3}x - 5\frac{1}{3}$

2) $y = x + 3$

3) $y = \frac{7}{3}x + \frac{7}{3}$

4) $y = 3x$

5) $y = 13$

6) $y = -8x + 5$

7) $y = \frac{3}{4}x + \frac{1}{4}$

8) $y = \frac{6}{5}x + \frac{21}{5}$

9) $y = -\frac{3}{5}x$

10) $y = -\frac{3}{4}x - 7$

11) $y = \frac{3}{2}x - 2$

12) $y = -3x - \frac{3}{4}$

13) $x = 3$

14) $y = -4$

15) $y = -3x - 4$

16) $y = \frac{1}{4}x + \frac{7}{10}$

17) $y = -\frac{2}{3}x - 4$

18) $y = -x - 2$

19) $y = -\frac{4}{5}x - \frac{18}{5}$

20) $y = -\frac{1}{7}x$

Equations of horizontal and vertical lines

1) $y = 2$

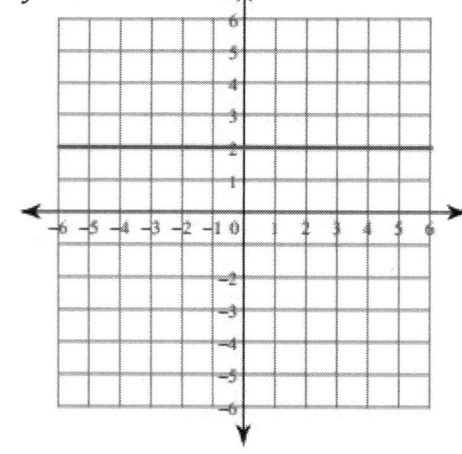

2) $y = 0$ (it is on x axes)

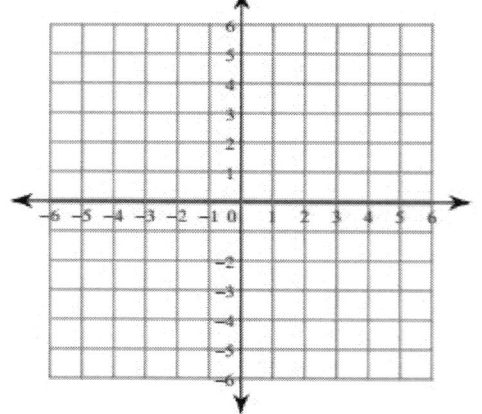

3) $x = 3$

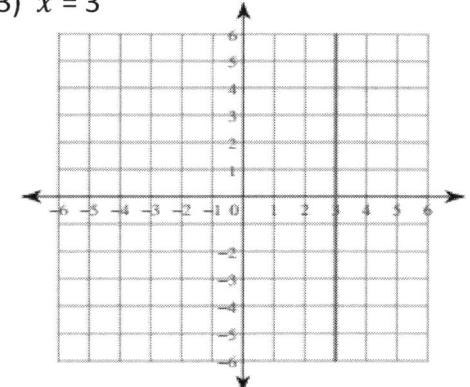

4) $x = -4$

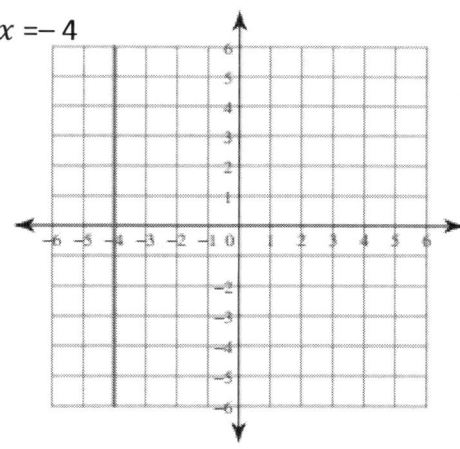

Function Notation

1) $v(t) = 7t$

2) $r(p) = 2p^2 +$
   $p - 1$

3) $h(g) = 14g + 7$

4) $f(x) = 3x - \frac{2}{5}$

5) 17

6) 0

7) -5

8) -71

9) 132

10) 8

11) -1

12) $2a - 4$

13) $-12b + 2$

14) $-2a - 7$

15) $x^6 + 2x^4 - 5$

16) $\frac{1}{4}x^2 + 2$

Adding and Subtracting Functions.

1) -4

2) $-3a^2 + a - 3$

3) 7

4) -1

5) -6

6) $a^2 + a + 1$

7) $-x^2 - 5x - 6$

8) $-t^2 + t - 8$

9) $2u^2 + 4u + 9$

10) $-t^2 + 10t - 20$

Multiplying and Dividing Functions

1) 88

2) $3x^4 - 7x^3 + 2x^2$

3) $-35\frac{1}{2}$

4) 0

5) -19

6) $x + 1$

7) $\frac{2}{a+1}$

8) $\frac{-2a+4}{a^3-1}$

Composition of functions

1) -7

2) -28

3) 23

4) 19

5) -6

6) 0

7) $-10 - 2a$

8) $100 - 12a$

9) $4a^2 + 20a - 22$

10) $-x + 5 - b$

11) $-x + 3 - b$

12) $-x^2 + 5 + b$

Chapter 10:

Quadratic Functions and Matrix

Topics that you'll learn in this chapter:

- ✓ Graphing Quadratic Functions
- ✓ Solving Quadratic Equations
- ✓ Use the Quadratic Formula and the Discriminant
- ✓ Solve Quadratic Inequalities
- ✓ Adding, Subtracting and Multiplications Matrices
- ✓ Finding Determinants of a Matrix
- ✓ Finding Inverse of a Matrix
- ✓ Matrix Equations

It's fine to work on any problem, so long as it generates interesting mathematics along the way – even if you don't solve it at the end of the day." – Andrew Wiles

Unlock Problems

❖ Graphing Quadratic Functions

✓ Quadratic functions in vertex form: $y = a(x - h)^2 + k$, vertex: (h, k)

✓ Quadratic functions in standard form: $y = ax^2 + bx + c$, vertex:(h, k), $h = \frac{-b}{2a}$; $k = ah^2 + bh + c$

✓ Step 1: Find the vertex of the quadratic function.

✓ Step 2: Plug in some values of x and solve for y. Then, find the points and graph the function.

❖ Solve a Quadratic Equation

✓ Write the equation in the form of Standard:

$ax^2 + bx + c = 0$

✓ Factorize the quadratic.

✓ Use quadratic formula if you couldn't factorize the quadratic.

✓ Quadratic formula: $x = \frac{-b \pm \sqrt{b^2 - 4ac}}{2a}$

Example: $x^2 + 5x = -6$

$x^2 + 5x + 6 = 0$

$(x + 3)(x + 2) = 0$

$(x + 3) = 0 \rightarrow x = -3$

$x + 2 = 0 \rightarrow x = -2$

❖ Use the Quadratic Formula and the Discriminant

✓ Discriminant determines the number of solutions for the given quadratic equation.

✓ The discriminant is the part of the quadratic formula under the square root.

$$x = \frac{-b \pm \sqrt{b^2 - 4ac}}{2a}$$

✓ If discriminant is positive, there are 2 solutions, if it's zero, there is one solution, and if it's negative, there is no solution for the quadratic function.

❖ Solve Quadratic Inequalities

✓ A quadratic inequality is one that can be written in one of the following standard forms:

- $ax^2 + bx + c > 0$
- $ax^2 + bx + c < 0$
- $ax^2 + bx + c \geq 0$
- $ax^2 + bx + c \leq 0$

✓ Solving quadratic inequality is similar to quadratic equations.

✓ In inequalities, we need to find a range of values of x that work in the inequality.

✓ Step 1: Solve the inequality and find the factors. (find the zeros)

✓ Step 2: Choose testing points to test each interval.

✓ Step 3: Determine the sign of the overall quadratic function in each interval.

❖ Adding and Subtracting Matrices

✓ We can add or subtract two matrices if they have the same dimensions.

✓ For addition or subtraction, add or subtract the corresponding entries, and place the result in the corresponding position in the resultant matrix.

❖ Matrix Multiplication

✓ Step 1: Make sure that it's possible to multiply the two matrices (the number of columns in the 1st one should be the same as the number of rows in the second one.)

✓ Step 2: The elements of each row of the first matrix should be multiplied by the elements of each column in the second matrix.

✓ Step 3: Add the products.

❖ Finding Determinants of a Matrix

✓ $\begin{bmatrix} a & b \\ c & d \end{bmatrix}$ $|A| = ad - bc$

✓ $\begin{bmatrix} a & b & c \\ d & e & f \\ g & h & i \end{bmatrix}$ $|A| = a(ei - fh) - b(di - fg) + c(dh - eg)$

❖ Finding Inverse of a Matrix

✓ $A = \begin{bmatrix} a & b \\ c & d \end{bmatrix}$ $A\text{-}1 = \frac{1}{|A|} \begin{bmatrix} d & -b \\ -c & a \end{bmatrix}$

❖ <u>Matrix Equations</u>

✓ In a matrix equation, a variable stand for a matrix.

✓ Matrix addition or scalar multiplication can be used to solve a matrix equation.

Graphing Quadratic Functions

Sketch the graph of each function. Identify the vertex and axis of symmetry.

1) $y = 2(x + 2)^2 - 3$

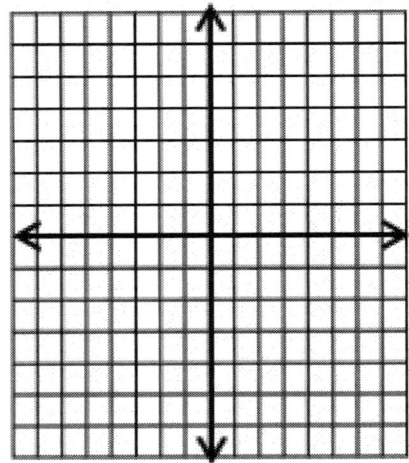

2) $y = -2(x + 2)^2 + 4$

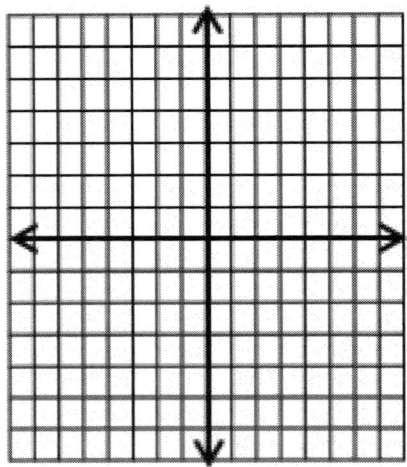

3) $x = -y^2 + 3y + 4$

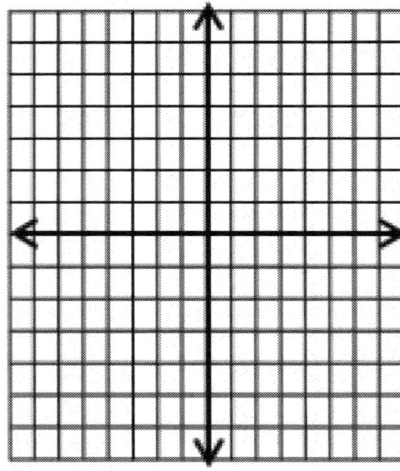

4) $y = (y + 1)^2 - 2$

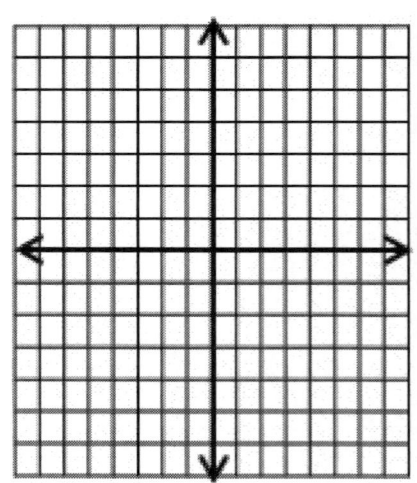

Solving Quadratic Equations

✏️ Solve each equation by factoring or by using the quadratic formula.

1) $x^2 + x - 20 = 2x$

2) $x^2 + 8x = -15$

3) $7x^2 - 14x = -7$

4) $6x^2 - 18x - 18 = 6$

5) $2x^2 + 6x - 24 = 12$

6) $2x^2 - 22x + 38 = -10$

7) $(2x + 5)(4x + 3) = 0$

8) $(x + 2)(x - 7) = 0$

9) $(x + 3)(x + 5) = 0$

10) $(5x + 7)(x + 4) = 0$

11) $-4x^2 - 8x - 3 = -3 - 5x^2$

12) $10x^2 = 27x - 18$

13) $7x^2 - 6x + 3 = 3$

14) $x^2 = 2x$

15) $2x^2 - 14 = -3x$

16) $10x^2 - 26x = -12$

17) $15x^2 + 80 = -80x$

18) $x^2 + 15x = -56$

Use the Quadratic Formula and the Discriminant

✎ **Find the value of the discriminant of each quadratic equation.**

1) $2x^2 + 5x - 4 = 0$

6) $6x^2 - 2x - 3 = 0$

2) $x^2 + 5x + 2 = 0$

7) $x(x - 1) = 0$

3) $5x^2 + x - 2 = 0$

8) $8x^2 - 9x = 0$

4) $-4x^2 - 4x + 5 = 0$

9) $3x^2 - 5x + 1 = 0$

5) $-2x^2 - x - 1 = 0$

10) $5x^2 + 6x + 4 = 0$

✎ **Find the discriminant of each quadratic equation then state the**

number of real and imaginary solution.

11) $8x^2 - 6x + 3 = 5x^2$

15) $4x^2 = 8x - 4$

12) $-4x^2 - 4x = 6$

16) $9x^2 + 6x + 6 = 5$

13) $-x^2 - 9 = 6x$

17) $9x^2 - 3x - 8 = -10$

14) $-9x^2 = -8x + 8$

18) $-2x^2 - 8x - 14 = -6$

Solve Quadratic Inequalities

✍ **Solve each quadratic inequality.**

1) $-x^2 - 5x + 6 > 0$

2) $x^2 - 5x - 6 < 0$

3) $x^2 + 4x - 5 > 0$

4) $x^2 - 2x - 3 \geq 0$

5) $x^2 - 1 < 0$

6) $17x^2 + 15x - 2 \geq 0$

7) $4x^2 + 20x - 11 < 0$

8) $12x^2 + 10x - 12 > 0$

9) $18x^2 + 23x + 5 \leq 0$

10) $-9x^2 + 29x - 6 \geq 0$

11) $-8x^2 + 6x - 1 \leq 0$

12) $5x^2 - 15x + 10 < 0$

13) $3x^2 - 5x \geq 4x^2 + 6$

14) $x^2 > 5x + 6$

15) $3x^2 + 7x \leq 5x^2 + 3x - 6$

16) $4x^2 - 12 > 3x^2 + x$

17) $3x^2 - 5x \geq 4x^2 + 6$

18) $2x^2 + 2x - 8 > x^2$

Adding and Subtracting Matrices

✎ Simplify.

1) $|-6 \quad 3 \quad -4| + |2 \quad -3 \quad -1|$

2) $\begin{vmatrix} 2 & 3 \\ -1 & -2 \\ -4 & -1 \end{vmatrix} + \begin{vmatrix} 0 & -1 \\ 1 & 0 \\ 2 & 5 \end{vmatrix}$

3) $\begin{vmatrix} -2 & 0 & -1 \\ 4 & -2 & 0 \end{vmatrix} - \begin{vmatrix} 6 & -2 & -1 \\ 1 & 4 & -3 \end{vmatrix}$

4) $|5 \quad 2| + |-2 \quad -7|$

5) $\begin{vmatrix} 1 \\ 4 \end{vmatrix} + \begin{vmatrix} 3 \\ 6 \end{vmatrix}$

6) $\begin{vmatrix} -r+t \\ -r \\ 3s \end{vmatrix} + \begin{vmatrix} r \\ -2t \\ -2r+2 \end{vmatrix}$

7) $\begin{vmatrix} z-2 \\ -4 \\ -1-5z \\ 2y \end{vmatrix} + \begin{vmatrix} -y \\ z \\ 5+z \\ 4z \end{vmatrix}$

8) $\begin{vmatrix} -4n & n+m \\ -2n & -4m \end{vmatrix} + \begin{vmatrix} 4 & -2 \\ m & 0 \end{vmatrix}$

9) $\begin{vmatrix} 2 & 3 \\ -6 & 5 \end{vmatrix} - \begin{vmatrix} 0 & -3 \\ 1 & 10 \end{vmatrix}$

10) $\begin{vmatrix} 1 & -5 & 9 \\ 4 & -3 & 11 \\ -6 & 3 & -15 \end{vmatrix} + \begin{vmatrix} 3 & 4 & -5 \\ 5 & 2 & 0 \\ 4 & -5 & 1 \end{vmatrix}$

Matrix Multiplication

✎ *Simplify.*

1) $\begin{vmatrix} -1 & -1 \\ -1 & 2 \end{vmatrix} \times \begin{vmatrix} -2 & -3 \\ 3 & 2 \end{vmatrix}$

2) $\begin{vmatrix} 0 & 3 \\ -3 & 1 \\ -5 & 1 \end{vmatrix} \times \begin{vmatrix} -2 & 2 \\ -2 & -4 \end{vmatrix}$

3) $\begin{vmatrix} 4 & 2 & 5 \\ 2 & 5 & 1 \end{vmatrix} \times \begin{vmatrix} 4 & 6 & -5 \\ 5 & -1 & 0 \end{vmatrix}$

4) $\begin{vmatrix} -4 \\ 0 \\ 2 \end{vmatrix} \times \begin{vmatrix} 2 & -1 \end{vmatrix}$

5) $\begin{vmatrix} 2 & -1 \\ 0 & 6 \\ -2 & -2 \end{vmatrix} \times \begin{vmatrix} -1 & 6 \\ 5 & 4 \end{vmatrix}$

6) $\begin{vmatrix} -1 & -3 \\ -2 & 3 \\ 3 & 0 \\ 4 & -2 \end{vmatrix} \times \begin{vmatrix} 1 & -2 & 1 \\ -1 & 0 & -3 \end{vmatrix}$

7) $\begin{vmatrix} -2 & -y \\ -x & -2 \end{vmatrix} \cdot \begin{vmatrix} -x & 0 \\ y & -2 \end{vmatrix}$

8) $\begin{vmatrix} 1 & -4v \end{vmatrix} \cdot \begin{vmatrix} -2u & -v \\ 0 & 3 \end{vmatrix}$

9) $\begin{vmatrix} -1 & 1 & -1 \\ 0 & 2 & -1 \\ 2 & -5 & 1 \\ -5 & 6 & 0 \end{vmatrix} \cdot \begin{vmatrix} 2 & 1 \\ 1 & -2 \\ 3 & 0 \end{vmatrix}$

10) $\begin{vmatrix} 5 & 3 & 5 \\ 1 & 5 & 0 \end{vmatrix} \cdot \begin{vmatrix} -4 & 2 \\ -3 & 4 \\ 3 & -5 \end{vmatrix}$

11) $\begin{vmatrix} -1 & 5 \\ -2 & 1 \end{vmatrix} \cdot \begin{vmatrix} 6 & -2 \\ 1 & 0 \end{vmatrix}$

12) $\begin{vmatrix} 0 & 2 \\ -2 & -5 \end{vmatrix} \cdot \begin{vmatrix} 2 & -1 \\ 3 & 0 \end{vmatrix}$

Finding Determinants of a Matrix

✍️ *Evaluate the determinant of each matrix.*

1) $\begin{vmatrix} 0 & -3 \\ -6 & -2 \end{vmatrix}$

2) $\begin{vmatrix} 0 & 3 \\ 2 & 6 \end{vmatrix}$

3) $\begin{vmatrix} -1 & 1 \\ -1 & 2 \end{vmatrix}$

4) $\begin{vmatrix} -2 & -9 \\ -1 & -10 \end{vmatrix}$

5) $\begin{vmatrix} -1 & 6 \\ 5 & 0 \end{vmatrix}$

6) $\begin{vmatrix} 8 & -6 \\ 0 & 9 \end{vmatrix}$

7) $\begin{vmatrix} 2 & -2 \\ 5 & -4 \end{vmatrix}$

8) $\begin{vmatrix} 2 & 6 \\ 3 & 9 \end{vmatrix}$

9) $\begin{vmatrix} 0 & 2 \\ -6 & 0 \end{vmatrix}$

10) $\begin{vmatrix} 0 & 4 \\ 4 & 5 \end{vmatrix}$

11) $\begin{vmatrix} 2 & -3 & 1 \\ 2 & 0 & -1 \\ 1 & 4 & 5 \end{vmatrix}$

12) $\begin{vmatrix} -5 & 0 & -1 \\ 1 & 2 & -1 \\ -3 & 4 & 1 \end{vmatrix}$

13) $\begin{vmatrix} 6 & 1 & 1 \\ 4 & -2 & 5 \\ 2 & 8 & 7 \end{vmatrix}$

14) $\begin{vmatrix} 3 & -5 & 3 \\ 2 & 1 & -1 \\ 1 & 0 & 4 \end{vmatrix}$

15) $\begin{vmatrix} 1 & 3 & 2 \\ 3 & -1 & -3 \\ 2 & 3 & 1 \end{vmatrix}$

Finding Inverse of a Matrix

✍ *Find the inverse of each matrix.*

1) $\begin{vmatrix} 4 & 7 \\ 2 & 6 \end{vmatrix}$

8) $\begin{vmatrix} -6 & -11 \\ 2 & 7 \end{vmatrix}$

2) $\begin{vmatrix} 2 & 1 \\ 3 & 2 \end{vmatrix}$

9) $\begin{vmatrix} -1 & 8 \\ -1 & 8 \end{vmatrix}$

3) $\begin{vmatrix} 4 & 3 \\ 2 & 1 \end{vmatrix}$

10) $\begin{vmatrix} -1 & 1 \\ 6 & 3 \end{vmatrix}$

4) $\begin{vmatrix} -9 & 6 \\ 4 & 3 \end{vmatrix}$

11) $\begin{vmatrix} 11 & 5 \\ 2 & 1 \end{vmatrix}$

5) $\begin{vmatrix} -3 & 2 \\ 1 & 3 \end{vmatrix}$

12) $\begin{vmatrix} 0 & 2 \\ 1 & 9 \end{vmatrix}$

6) $\begin{vmatrix} 2 & 4 \\ 5 & 2 \end{vmatrix}$

13) $\begin{vmatrix} 0 & 0 \\ -6 & 3 \end{vmatrix}$

7) $\begin{vmatrix} 0 & 7 \\ 3 & 2 \end{vmatrix}$

14) $\begin{vmatrix} 3 & 4 \\ 6 & 8 \end{vmatrix}$

Matrix Equations

✐ *Solve each equation.*

1) $\begin{vmatrix} -1 & 2 \\ -2 & 5 \end{vmatrix} z = \begin{vmatrix} 6 \\ 20 \end{vmatrix}$

2) $2x = \begin{vmatrix} 12 & -12 \\ 24 & -8 \end{vmatrix}$

3) $\begin{vmatrix} -3 & 2 \\ -11 & 6 \end{vmatrix} = \begin{vmatrix} 2 & 8 \\ 5 & 9 \end{vmatrix} - x$

4) $Y - \begin{vmatrix} -2 \\ -4 \\ 10 \\ 10 \end{vmatrix} = \begin{vmatrix} -6 \\ 6 \\ -16 \\ 0 \end{vmatrix}$

5) $\begin{vmatrix} -1 & -3 \\ 0 & -4 \end{vmatrix} C = \begin{vmatrix} 10 \\ 8 \end{vmatrix}$

6) $\begin{vmatrix} -1 & -3 \\ 2 & 8 \end{vmatrix} B = \begin{vmatrix} -8 & -2 & 8 \\ 22 & 0 & -20 \end{vmatrix}$

7) $\begin{vmatrix} -1 & 1 \\ 5 & -2 \end{vmatrix} C = \begin{vmatrix} 1 \\ -11 \end{vmatrix}$

8) $\begin{vmatrix} 1 & 2 \\ 3 & 4 \end{vmatrix} C = \begin{vmatrix} 11 \\ 21 \end{vmatrix}$

9) $\begin{vmatrix} 0 & -4 \\ 3 & 3 \end{vmatrix} Z = \begin{vmatrix} 20 \\ 6 \end{vmatrix}$

10) $\begin{vmatrix} -10 \\ 15 \\ -20 \end{vmatrix} = 5B$

11) $\begin{vmatrix} -10 \\ 4 \\ 3 \end{vmatrix} = y - \begin{vmatrix} 7 \\ -5 \\ -11 \end{vmatrix}$

12) $-4b - \begin{vmatrix} 5 \\ 2 \\ -6 \end{vmatrix} = \begin{vmatrix} -33 \\ -2 \\ -22 \end{vmatrix}$

Answers of Worksheets – Chapter 10

Graphing quadratic functions

1)

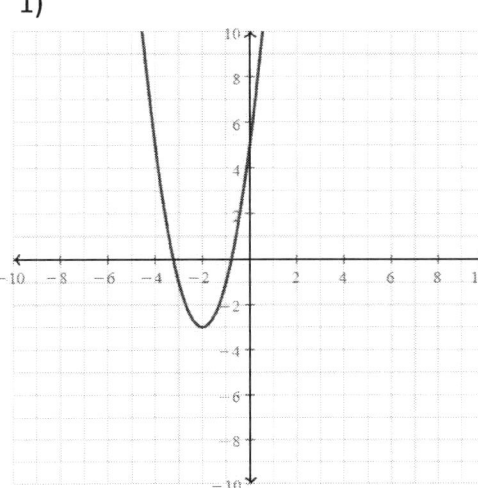

2)

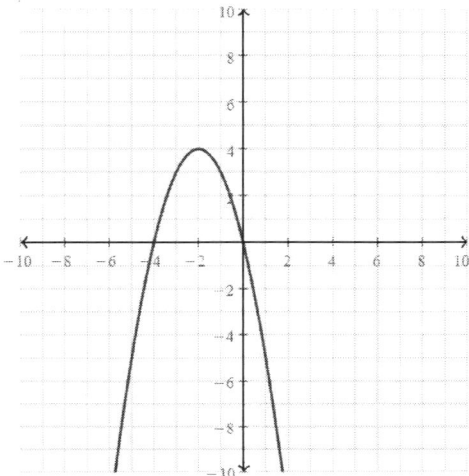

3)

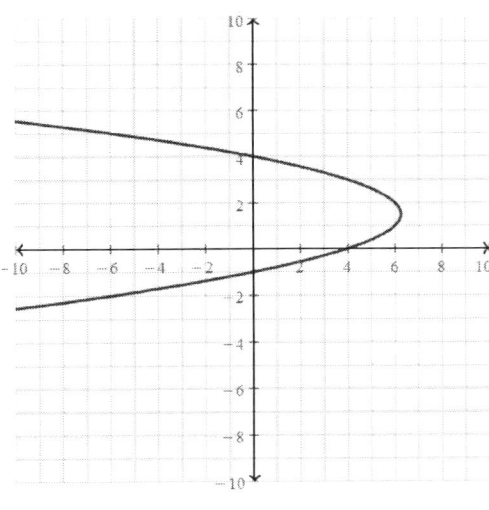

4)

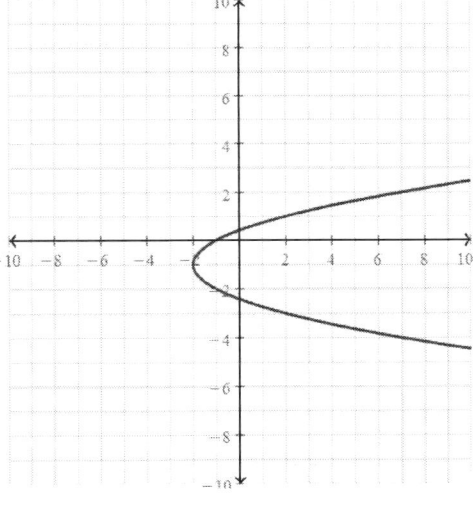

Solving quadratic equations

1) {5, − 4}

2) {−5, −3}

3) {1}

4) {4, −1}

5) {3, −6}

6) {3, 8}

7) $\{-\frac{5}{2}, -\frac{3}{4}\}$

8) {− 2, 7}

9) {− 3, − 5}

10) $\{-\frac{7}{5}, -4\}$

11) {8, 0}

12) $\{\frac{6}{5}, \frac{3}{2}\}$

13) $\{\frac{6}{7}, 0\}$

14) $\{2, 0\}$

15) $\{-\frac{7}{2}, 2\}$

16) $\{\frac{3}{5}, 2\}$

17) $\{-\frac{4}{3}, -4\}$

18) $\{-8, -7\}$

Use the quadratic formula and the discriminant

1) 57

2) 17

3) 41

4) 96

5) −7

6) 76

7) 1

8) 81

9) 13

10) −44

11) 0, one real solution

12) −80, no solution

13) 0, one real solution

14) −224, no solution

15) 0, one real solution

16) 0, one real solution

17) −63, two imaginary solution

18) 0, one real solution

Solve quadratic inequalities

1) $-6 < x < 1$

2) $-1 < x < 6$

3) $x < -5$ or $x > 1$

4) $x \le -1$ or $x \ge 3$

5) $-1 < x < 1$

6) $x \le -1$ or $x \ge \frac{2}{17}$

7) $-\frac{11}{2} < x < \frac{1}{2}$

8) $x < -\frac{3}{2}$ or $x > \frac{2}{3}$

9) $-1 \le x \le -\frac{5}{18}$

10) $\frac{2}{9} \le x \le 3$

11) $x \le \frac{1}{4}$ or $x \ge \frac{1}{2}$

12) $1 < x < 2$

13) $-3 \le x \le -2$

14) $x < -1$ or $x > 6$

15) $x \le -1$ or $x \ge 3$

16) $x < -3$ or $x > 4$

17) $-3 \le x \le -2$

18) $x < -4$ or $x > 2$

Adding and Subtracting Matrices

1) $|-4 \quad 0 \quad -5|$

2) $\begin{vmatrix} 2 & 2 \\ 0 & -2 \\ -2 & 4 \end{vmatrix}$

3) $\begin{vmatrix} -8 & 2 & 0 \\ 3 & -6 & 3 \end{vmatrix}$

4) $|3 \quad -5|$

5) $\begin{vmatrix} 4 \\ 10 \end{vmatrix}$

6) $\begin{vmatrix} t \\ -r - 2t \\ 3s - 2r + 2 \end{vmatrix}$

7) $\begin{vmatrix} z - 2 - y \\ -4 + z \\ 4 - 4z \\ 2y + 4z \end{vmatrix}$

8) $\begin{vmatrix} -4n + 4 & n + m - 2 \\ -2n + m & -4m \end{vmatrix}$

9) $\begin{vmatrix} 2 & 6 \\ -7 & -5 \end{vmatrix}$

10) $\begin{vmatrix} 4 & -1 & 4 \\ 9 & -1 & 11 \\ -2 & -2 & -14 \end{vmatrix}$

Matrix Multiplication

1) $\begin{vmatrix} -1 & 1 \\ 8 & 7 \end{vmatrix}$

2) $\begin{vmatrix} -6 & -12 \\ 4 & -10 \\ 8 & -14 \end{vmatrix}$

3) Undefined

4) $\begin{vmatrix} -8 & 4 \\ 0 & 0 \\ 4 & -2 \end{vmatrix}$

5) $\begin{vmatrix} -7 & 8 \\ 30 & 24 \\ -8 & -20 \end{vmatrix}$

6) $\begin{vmatrix} 2 & 2 & 8 \\ -5 & 4 & -11 \\ 3 & -6 & 3 \\ 6 & -8 & 10 \end{vmatrix}$

7) $\begin{vmatrix} 2x - y^2 & 0 \\ x^2 - 2y & 4 \end{vmatrix}$

8) $\begin{vmatrix} -2u & -13v \end{vmatrix}$

9) $\begin{vmatrix} -4 & -3 \\ -1 & -4 \\ 2 & 12 \\ -4 & -17 \end{vmatrix}$

10) $\begin{vmatrix} -14 & -3 \\ -19 & 22 \end{vmatrix}$

11) $\begin{vmatrix} -1 & 2 \\ -11 & 4 \end{vmatrix}$

12) $\begin{vmatrix} 6 & 0 \\ -19 & 2 \end{vmatrix}$

Finding Determinants of a Matrix

1) −18

2) -6

3) −1

4) 11

5) −30

6) 72

7) 2

8) 0

9) 12

10) −16

11) 49

12) -40

13) −30

14) 54

15) 3

Finding Inverse of a Matrix

1) $\begin{vmatrix} \dfrac{6}{10} & \dfrac{-7}{10} \\ \dfrac{-1}{5} & \dfrac{2}{5} \end{vmatrix}$

2) $\begin{vmatrix} 2 & -1 \\ -3 & 2 \end{vmatrix}$

3) $\begin{vmatrix} -\dfrac{1}{2} & \dfrac{3}{2} \\ 1 & -2 \end{vmatrix}$

4) $\begin{vmatrix} -\dfrac{3}{51} & \dfrac{6}{51} \\ \dfrac{4}{51} & \dfrac{9}{51} \end{vmatrix}$

5) $\begin{vmatrix} -\dfrac{3}{11} & \dfrac{2}{11} \\ \dfrac{1}{11} & \dfrac{3}{11} \end{vmatrix}$

6) $\begin{vmatrix} -\dfrac{2}{16} & \dfrac{4}{16} \\ \dfrac{5}{16} & -\dfrac{2}{16} \end{vmatrix}$

7) $\begin{vmatrix} -\dfrac{2}{21} & \dfrac{7}{21} \\ \dfrac{3}{21} & 0 \end{vmatrix}$

8) $\begin{vmatrix} -\dfrac{7}{20} & -\dfrac{11}{20} \\ \dfrac{2}{20} & \dfrac{6}{20} \end{vmatrix}$

9) No inverse exists

10) $\begin{vmatrix} -\dfrac{3}{9} & \dfrac{1}{9} \\ \dfrac{6}{9} & \dfrac{1}{9} \end{vmatrix}$

11) $\begin{vmatrix} 1 & -5 \\ -2 & 11 \end{vmatrix}$

12) $\begin{vmatrix} -\dfrac{9}{2} & 1 \\ \dfrac{1}{2} & 0 \end{vmatrix}$

13) No inverse exists

14) No inverse exists

Matrix Equations

1) $\begin{vmatrix} 10 \\ 8 \end{vmatrix}$

2) $\begin{vmatrix} 6 & -6 \\ 12 & -4 \end{vmatrix}$

3) $\begin{vmatrix} 5 & 6 \\ 16 & 3 \end{vmatrix}$

4) $\begin{vmatrix} -4 \\ 2 \\ -6 \\ 10 \end{vmatrix}$

5) $\begin{vmatrix} -4 \\ -2 \end{vmatrix}$

6) $\begin{vmatrix} -1 & 8 & -2 \\ 3 & -2 & -2 \end{vmatrix}$

7) $\begin{vmatrix} -3 \\ -2 \end{vmatrix}$

8) $\begin{vmatrix} -1 \\ 6 \end{vmatrix}$

9) $\begin{vmatrix} 7 \\ -5 \end{vmatrix}$

10) $\begin{vmatrix} -2 \\ 3 \\ -4 \end{vmatrix}$

11) $\begin{vmatrix} -3 \\ -1 \\ -8 \end{vmatrix}$

12) $\begin{vmatrix} 7 \\ 0 \\ 7 \end{vmatrix}$

Chapter 11: Logarithms

Topics that you'll learn in this chapter:

✓ Rewriting Logarithms

✓ Evaluating Logarithms

✓ Properties of Logarithms

✓ Natural Logarithms

✓ Solving Exponential Equations Requiring Logarithms

✓ Solving Logarithmic Equations

Mathematics is an art of human understanding. – William Thurston

Unlock problems

❖ Rewriting Logarithms

✓ $\log_b y = x$ is equivalent to $y = b^x$

Example:

$$\log_4 y = 3 \iff y = 4^3$$

❖ Evaluating Logarithms

✓ Change of Base Formula: $\log_b (x) = \dfrac{\log_d (x)}{\log_d (b)}$

❖ Properties of Logarithms

✓ $a^{\log_a b} = b$

✓ $\log_a 1 = 0$

✓ $\log_a a = 1$

✓ $\log_a (x \cdot y) = \log_a x + \log_a y$

✓ $\log_a \dfrac{x}{y} = \log_a x - \log_a y$

Example:

$$\log_a \frac{1}{x} = -\log_a x$$

$$\log_a x^p = p \log_a x$$

$$\log_{x^k} x = \frac{1}{x} \log_a x, \text{ for } k \neq 0$$

$$\log_a x = \log_{a^c} x^c$$

$$\log_a x = \frac{1}{\log_x a}$$

❖ Natural Logarithms

✓ $\ln(x \cdot y) = \ln(x) + \ln(y)$

✓ $\ln\left(\frac{x}{y}\right) = \ln(x) - \ln(y)$

✓ $\ln(x^y) = y \cdot \ln(x)$

❖ Solving Exponential Equations Requiring Logarithms

✓ if $b^m = b^n$ then $m = n$

❖ Solving Logarithmic Equations

✓ Convert the logarithmic equation to an exponential equation when it's possible. (If no base is indicated, the base of the logarithm is 10)

✓ Condense logarithms if you have more than one log on one side of the equation.

✓ Plug in the answers back into the original equation and check to see the solution works.

Rewriting Logarithms

✎ Rewrite each equation in exponential form.

1) $\log_{0.81} 0.9 = \frac{1}{2}$

2) $\log_{100} 1000 = 1.50$

3) $\log_8 64 = 2$

4) $\log_{10} 100 = 2$

✎ Rewrite each equation in exponential form.

5) $\log_a \frac{6}{7} = b$

6) $\log_x y = 8$

7) $\log_{15} n = m$

8) $\log_y x = -9$

9) $\log_a b = 33$

10) $\log_{\frac{1}{3}} v = u$

✎ Evaluate each expression.

11) $\log_3 27$

12) $\log_6 216$

13) $\log_5 25$

14) $\log_8 4$

Evaluating Logarithms

✎ Evaluate each expression.

1) $\log_2 16$

2) $\log_3 81$

3) $\log_2 8$

4) $\log_3 9$

5) $\log_9 81$

6) $\log_6 \frac{1}{36}$

7) $\log_{27} \frac{1}{3}$

8) $\log_{80} 300$

9) $\log_5 \frac{1}{125}$

10) $\log_4 256$

11) $\log_7 343$

12) $\log_3 \frac{1}{27}$

13) $\log_9 729$

14) $\log_8 4096$

Properties of Logarithms

✎ **Expand each logarithm.**

1) $\log \left(\frac{4}{5}\right)^3$

2) $\log (7.3^3)$

3) $\log \left(\frac{5}{7}\right)^2$

4) $\log \dfrac{3^3}{5}$

5) $\log (x \cdot y)^7$

6) $\log (3 \cdot 8)$

7) $\log (2 \cdot 5)$

8) $\log (x^4 \cdot y \cdot z^5)$

9) $\log \dfrac{u^3}{v}$

10) $\log \dfrac{x}{y^9}$

✎ **Condense each expression to a single logarithm.**

11) $\log 4 - \log 7$

12) $4 \log 3 - 3 \log 2$

13) $\log 5 - 3 \log 11$

14) $5 \log_7 a + 9 \log_7 b$

15) $2\log_2 x - 5 \log_2 y$

16) $\log_6 u - 7 \log_6 v$

17) $2 \log_6 u + 5 \log_6 v$

18) $6 \log_5 u - 10 \log_5 v$

Natural Logarithms

✎ **Solve.**

1) $e^x = 2$

2) $\ln (\ln x) = 7$

3) $e^x = 6$

4) $\ln (5x + 6) = 4$

5) $\ln (9x - 1) = 1$

6) $\ln x = \dfrac{1}{3}$

7) $x = e^{\frac{1}{2}}$

8) $\ln x = \ln 3 + \ln 8$

✍ **Evaluate without using a calculator.**

9) $\ln\sqrt{e}$

10) $\ln e^5$

11) $6 \ln e$

12) $\ln\left(\frac{1}{e}\right)$

13) $e^{\ln 12}$

14) $e^{3\ln 3}$

15) $e^{4\ln 2}$

16) $\ln 1$

Solving Exponential Equations Requiring Logarithms

✍ **Solve each equation.**

1) $2^{r+1} = 1$

2) $216^x = 36$

3) $4^{-3v-3} = 16$

4) $4^{2n} = 16$

5) $\frac{216^{2a}}{36^{-a}} = 216$

6) $24. \ 24^{-v} = 576$

7) $2^{2n} = 4$

8) $\left(\frac{1}{7}\right)^n = 49$

9) $32^{2x} = 4$

10) $2^{-2x} = 2^{x-1}$

11) $2^{2n} = 64$

12) $6^{3n} = 216$

13) $4^{-2k} = 256$

14) $6^{2r} = 6^{3r}$

15) $10^{5x} = 10000$

16) $36. \ 6^{-v} = 216$

17) $\frac{64}{16^{-3m}} = 16^{-2m-2}$

18) $3^{-2n} \cdot 3^{n+1} = 3^{-2n}$

Solving Logarithmic Equations

✍ **Solve each equation.**

1) $2 \log_6 - 2x = 0$

2) $- \log_2 3x = 2$

3) $\log x + 6 = 3$

4) $\log x - \log 2 = 1$

5) $\log x + \log 2 = 2$

6) $\log 5 + \log x = 1$

7) $\log x + \log 8 = \log 16$

8) $- 2 \log_2 (x - 3) = - 10$

9) $\log 7x = \log (x + 6)$

10) $\log(9k - 5) = \log(3k - 1)$

11) $\log(5p - 1) = \log(- 4p + 6)$

12) $- 10 + \log_3(n + 2) = - 10$

13) $\log_8(12x + 3) = \log_8(x^2 + 30)$

14) $\log_{12}(v^2 - 38) = \log_{12}(- 5v - 2)$

15) $\log(12 + 2b) = \log(b^2 - 2b)$

16) $\log_8(x + 4) - \log_8 x = \log_8 2$

17) $\log_2 4 + \log_2 x^2 = \log_2 36$

18) $\log_8(x + 1) - \log_8 x = \log_8 32$

Answers of Worksheets – Chapter 11

Rewriting Logarithms

1) $0.81^{\frac{1}{2}} = 0.9$

2) $100^{1.5} = 1000$

3) $8^2 = 64$

4) $10^2 = 100$

5) $a^b = \frac{6}{7}$

6) $x^8 = y$

7) $15^m = n$

8) $y^{-9} = x$

9) $a^{33} = b$

10) $(\frac{1}{3})^u = v$

11) 3

12) 3

13) 2

14) $\frac{2}{3}$

Evaluating Logarithms

1) 4

2) 4

3) 3

4) 2

5) 2

6) -2

7) $-\frac{1}{3}$

8) 1.3

9) -3

10) 4

11) 3

12) -3

13) 3

14) 4

Properties of Logarithms

1) $3 \log 4 - 3 \log 5$

2) $\log 7 + 3 \log 3$

3) $2\log 5 - 2 \log 7$

4) $3 \log 3 - \log 5$

5) $7 \log x + 7 \log y$

6) $\text{Log } 3 + \log 8$

7) $\text{Log } 2 + \log 5$

8) $4\text{Log } x + \log y + 5 \log z$

9) $3 \log u - \log v$

10) $\text{Log } x - 9 \log y$

11) $\log \frac{4}{7}$

12) $\log \frac{3^4}{2^3}$

13) $\log \frac{5}{11^3}$

14) $\log_7 (a^5 b^9)$

15) $\log_2 \frac{x^2}{y^5}$

16) $\log_6 \frac{u}{v^7}$

17) $\log_6 (v^5 u^2)$

18) $\log_5 \frac{u^6}{v^{10}}$

Natural Logarithms

1) $x = \ln 2$

2) $x = e^{e^7}$

3) $x = \ln 6$

4) $x = \dfrac{e^4 - 6}{5}$

5) $x = \dfrac{e + 1}{9}$

6) $x = e^{\frac{1}{3}}$

7) $\ln x = \dfrac{1}{2}$

8) $x = 24$

9) $\dfrac{1}{2}$

10) 5

11) 6

12) -1

13) 12

14) 27

15) 16

16) 0

Solving Exponential Equations Requiring Logarithms

1) -1

2) $\dfrac{1}{12}$

3) $-\dfrac{5}{3}$

4) 1

5) $\{\dfrac{3}{8}\}$

6) -1

7) 1

8) -2

9) $\dfrac{1}{5}$

10) $\dfrac{1}{3}$

11) 3

12) 1

13) -2

14) 0

15) $\dfrac{4}{5}$

16) -1

17) $-\dfrac{7}{10}$

18) -1

Solving Logarithmic Equations

1) $\{-\dfrac{1}{2}\}$

2) $\{12\}$

3) $\{\dfrac{1}{1000}\}$

4) $\{20\}$

5) $\{50\}$

6) $\{2\}$

7) $\{2\}$

8) $\{35\}$

9) $\{1\}$

10) $\{\dfrac{2}{3}\}$

11) $\{\dfrac{7}{9}\}$

12) $\{-1\}$

13) $\{9, 3\}$

14) $\{-9, 4\}$

15) $\{6, -2\}$

16) $\{4\}$

17) $\{3, -3\}$

18) $\{\dfrac{1}{31}\}$

Chapter 12: Geometry

Topics that you'll learn in this chapter:

✓ The Pythagorean Theorem

✓ Area of Triangles and Trapezoids

✓ Area and Circumference of Circles

✓ Area and Perimeter of Polygons

✓ Area of Squares, Rectangles, and Parallelograms

✓ Volume of Cubes, Rectangle Prisms, and Cylinder

✓ Surface Area of Cubes, Rectangle Prisms, and Cylinder

Mathematics is, as it were, a sensuous logic, and relates to philosophy as do the arts, music, and plastic art to poetry. — K. Shegel

Unlock problems

❖ The Pythagorean Theorem

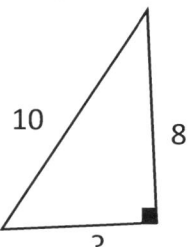

✓ In any right triangle:

$$a^2 + b^2 = c^2$$

Example:

Missing side $= 10^2 - 8^2 = 100 - 64 = 36$

$? = 6$

❖ Angles

✓ **Adjacent:** Two angles that have a common side, a common

vertex, and don't overlap.

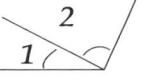

✓ **Vertical:** Two angles share same vertex.

✓ **Complementary:** Sum of the measure of two complementary

angles are 90°

✓ **Supplementary:** Sum of the measure of two supplementary

angles are 180°

❖ Area of Triangles

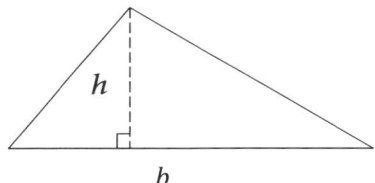

✓ Area $= \frac{1}{2} (base \times height)$

❖ Area of Trapezoids

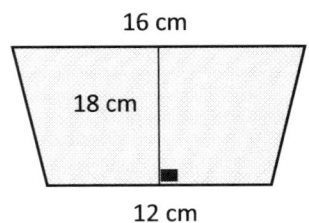

16 cm

18 cm

12 cm

✓ $A = \frac{1}{2}h(b_1 + b_2)$

Example: $A = 252\ cm^2$

❖ Area of Polygons

✓ Area of Rectangles = Length × width

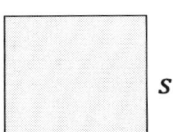

✓ Area of Squares= s^2

s

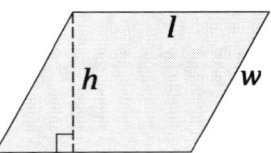

l

h

w

✓ Area of Parallelograms = length × height

20

Example:

$Area = 220$

11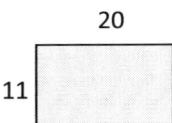

❖ Perimeter of Polygons

✓ Perimeter of a square = $4s$

s

✓ Perimeter of a rectangle= $2(l + w)$

w

l

✓ Perimeter of Pentagon = $6a$

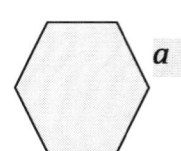

a

✓ Perimeter of a parallelogram $= 2(l + w)$

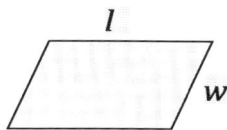

✓ Perimeter of trapezoid $= a + b + c + d$

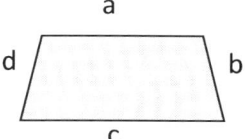

❖ Area and Circumference of Circles

✓ Area $= \pi r^2$

✓ Circumference $= 2\pi r$

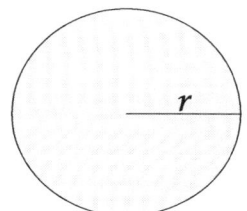

Example:

If the radius of a circle is 3, then:

$Area = 28.27$

Circumference $= 18.85$

❖ Volume of Cubes

✓ Volume is the measure of the amount of space inside of a solid figure, like a cube, ball, cylinder or pyramid.

✓ Volume of a cube $= (one\ side)^3$

✓ Volume of a rectangle prism: $Length \times Width \times Height$

❖ Volume of Rectangle Prisms

✓ Volume of rectangle prism: $length \times width \times height$

Example:

$10 \times 5 \times 8 = 400 m^3$

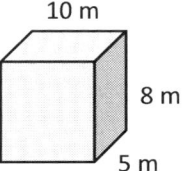

❖ Surface Area of Cubes

✓ Surface Area of a cube $= 6 \times (one\ side\ of\ the\ cube)^2$

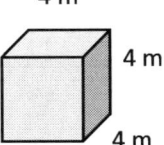

4 m

4 m

4 m

Example:

$6 \times 4^2 = 96m^2$

❖ Surface Area of a Rectangle Prism

✓ Surface Area of a Rectangle Prism Formula:

$SA = 2[(width \times length) + (height \times length) + width \times height)]$

❖ Volume of a Cylinder

✓ Volume of Cylinder Formula $= \pi(radius)^2 \times height$

$\pi = 3.14$

❖ Surface Area of a Cylinder

✓ Surface area of a cylinder: $SA = 2\pi r^2 + 2\pi rh$

Example:

Surface area $= 1727$

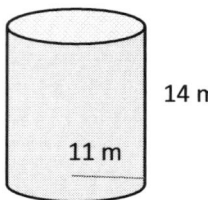

14 m

11 m

❖ Volume of Pyramids and Cones

✓ Volume of a pyramid: $V = \frac{1}{3}b.h$

✓ Volume of a cone: $V = \frac{1}{3}\pi r^2 h$

❖ <u>Surface Area of Pyramids and Cones</u>

✓ Surface Area Pyramid:

$$SA = lw + l\sqrt{(\frac{w}{2})^2 + h^2} + w\sqrt{(\frac{l}{2})^2 + h^2}$$

✓ Surface Area Cone:

$$SA = \pi r\,(r + \sqrt{h^2 + r^2})$$

The Pythagorean Theorem

✍ Do the following lengths form a right triangle?

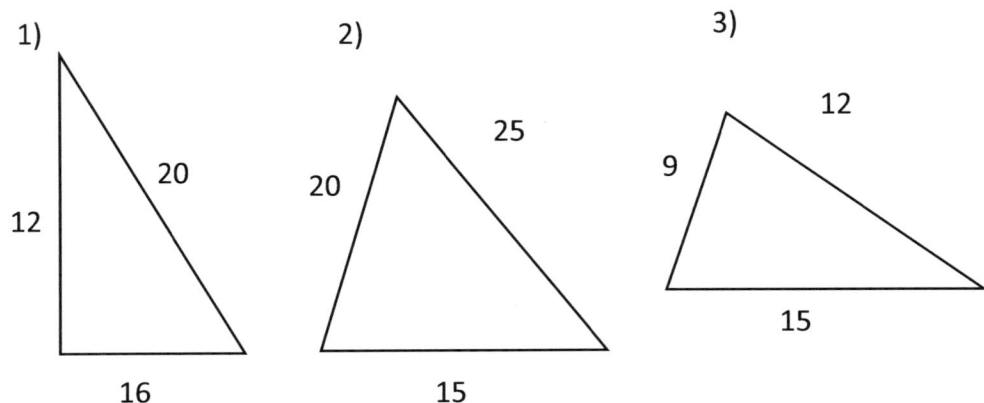

1)
12
20
16

2)
25
20
15

3)
12
9
15

✍ Find each missing length to the nearest tenth.

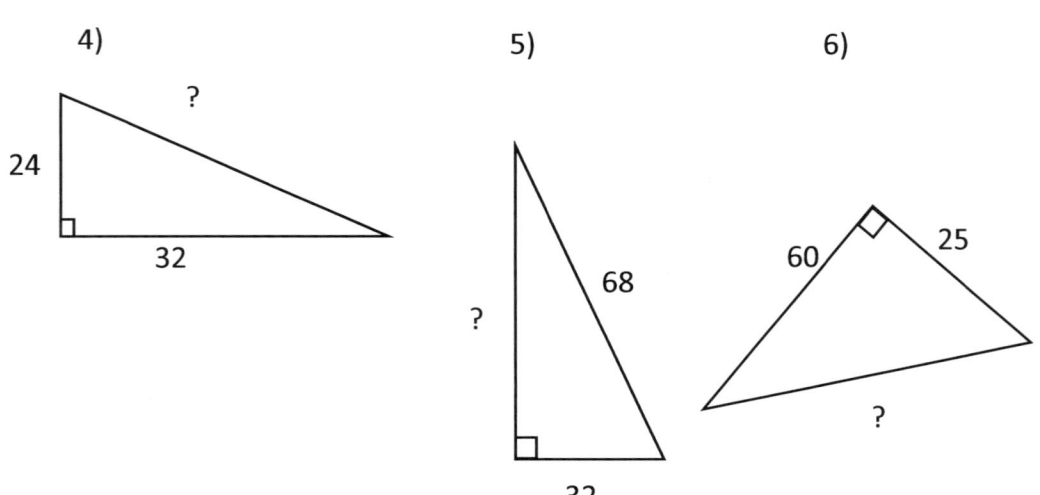

4)
?
24
32

5)
?
68
32

6)
60
25
?

Angles

✎ **What is the value of x in the following figures?**

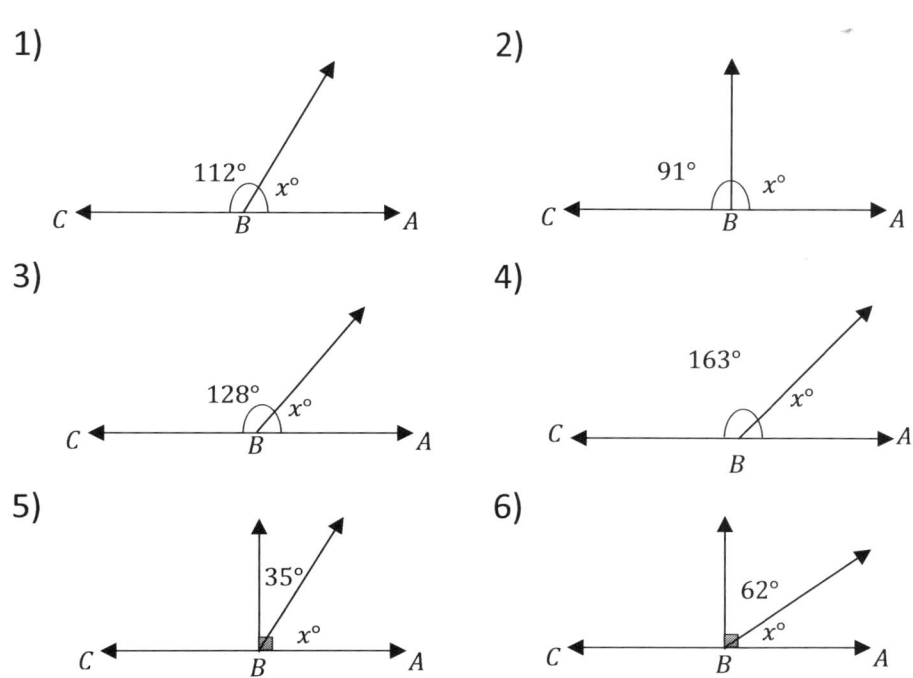

1)

$112°$ $x°$

2)

$91°$ $x°$

3)

$128°$ $x°$

4)

$163°$ $x°$

5)

$35°$ $x°$

6)

$62°$ $x°$

✎ *Solve.*

7) Two complementary angles have equal measures. What is the measure of each angle? _____

8) The measure of an angle is two third the measure of its supplement. What is the measure of the angle? _____

Area of Triangles

✎ **Find the area of each.**

1)

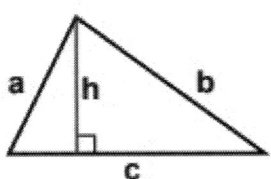

c = 12 mi

h = 4.5 mi

2)

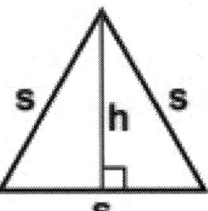

s = 8 m

h = 9.4 m

3)

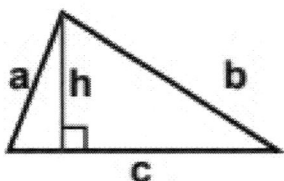

a = 4 m

b = 11 m

c = 16 m

h = 13.6 m

4)

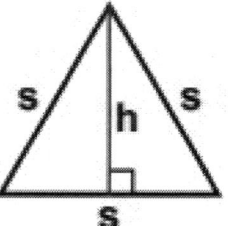

s = 6 m

h = 6.71 m

Area of Trapezoids

✎ **Calculate the area for each trapezoid.**

1)

11 cm

7 cm

13 cm

2)

18 m

12 m

24 m

3)

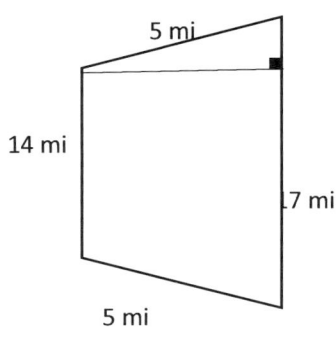

5 mi

14 mi

7 mi

5 mi

4)

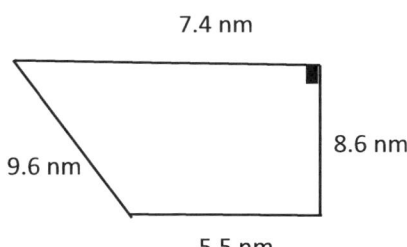

7.4 nm

9.6 nm

8.6 nm

5.5 nm

Area and Perimeter of Polygons

✎ Find the area and perimeter of each

1)

18 yd

27.5 yd

27.5 yd

18 yd

2)

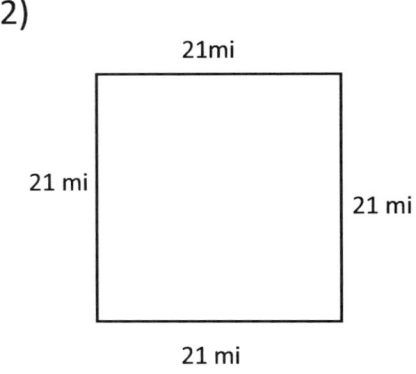

21mi

21 mi

21 mi

21 mi

3)

15.2 ft

12.8 ft

10 ft

12.8 ft

15.2 ft

4)

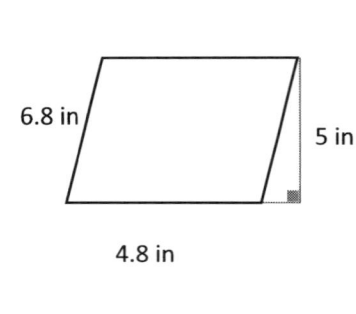

6.8 in

5 in

4.8 in

5)

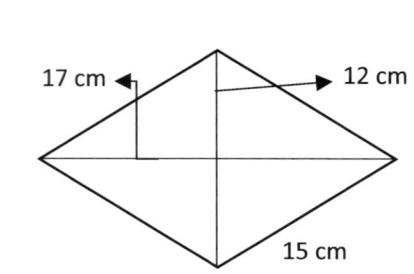

17 cm

12 cm

15 cm

6)

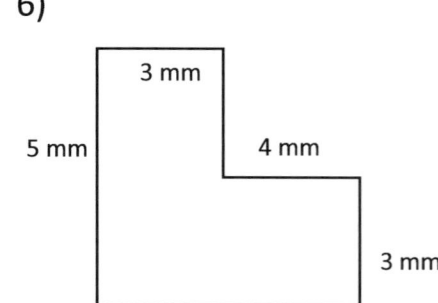

3 mm

5 mm

4 mm

3 mm

✏️ **Find the perimeter of each shape.**

7)

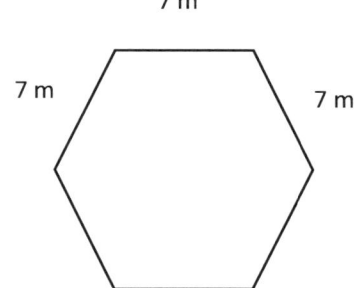

7 m

7 m

7 m

8)

12 mm

12 mm

9)

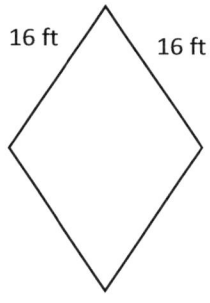

16 ft

16 ft

10)

21 in

17 in

11)

14 cm

12)

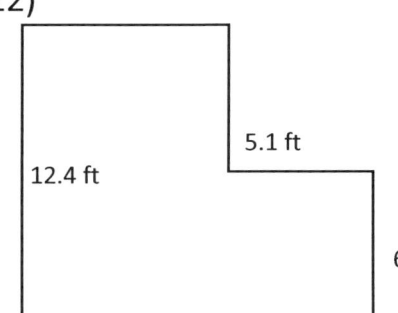

5.1 ft

12.4 ft

6.2 ft

17.1 ft

Area and Circumference of Circles

✏️ **Find the area and circumference of each.** ($\pi = 3.14$)

1)

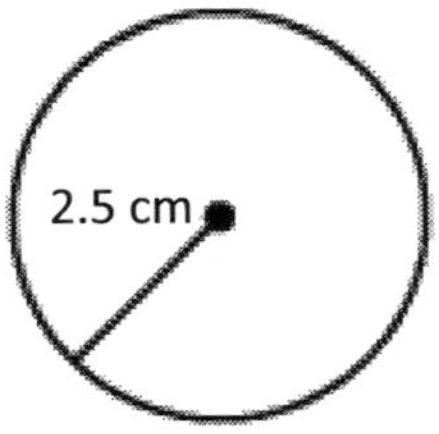

2.5 cm

2)

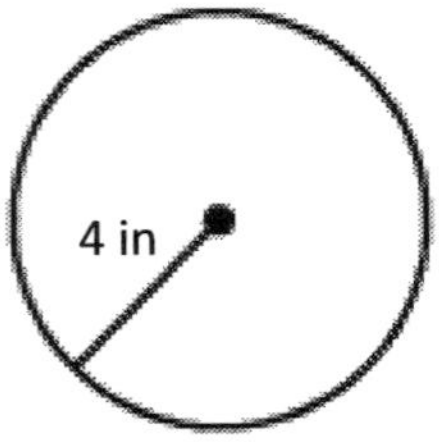

4 in

3)

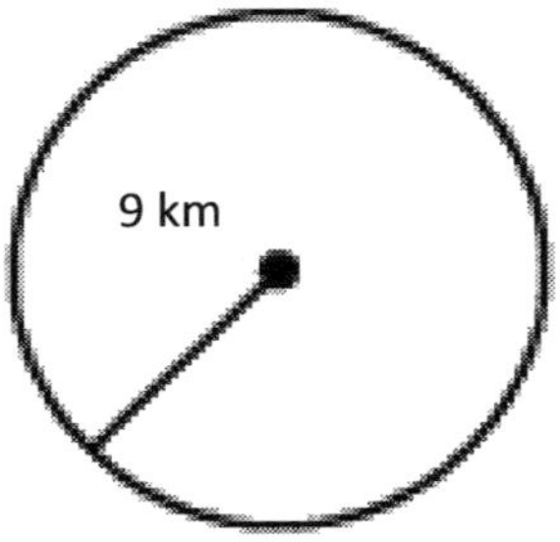

9 km

4)

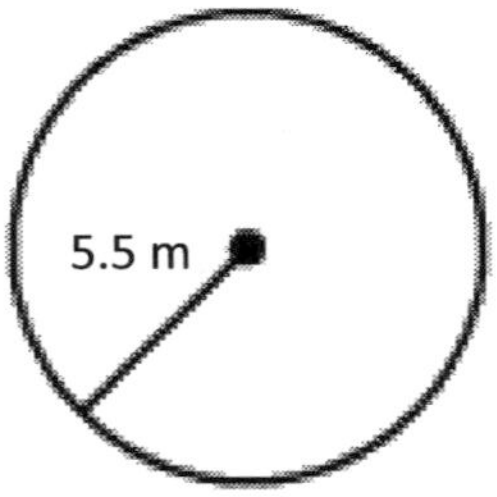

5.5 m

5)

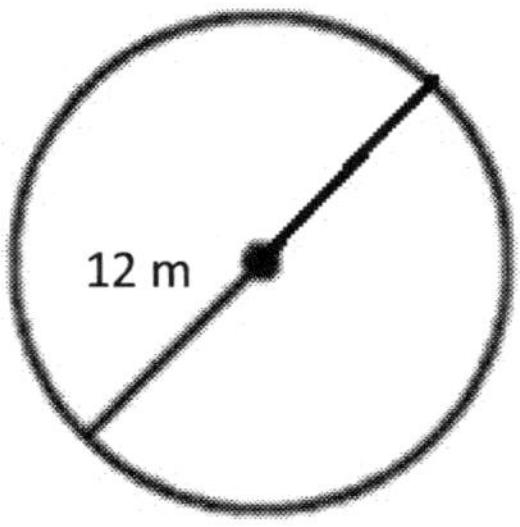

12 m

6)

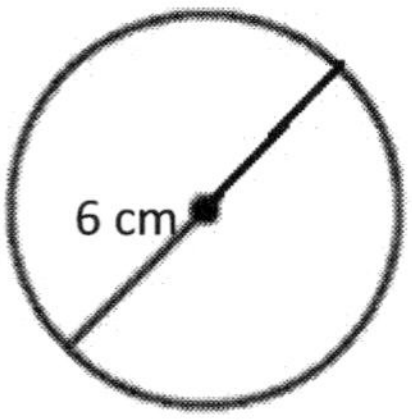

6 cm

7)

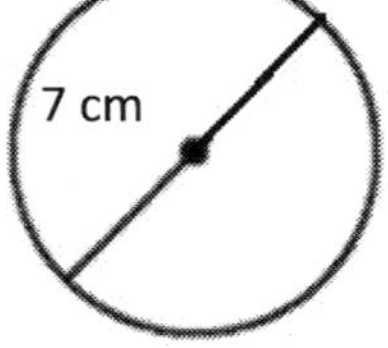

7 cm

8)

3 in

Volume of Cubes

✎ **Find the volume of each.**

1)

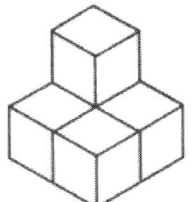

2)

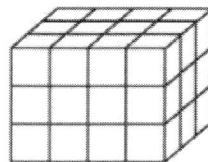

3)

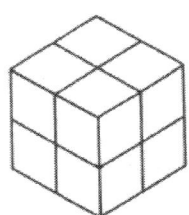

4)

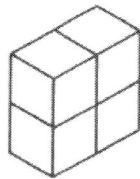

5)

6)

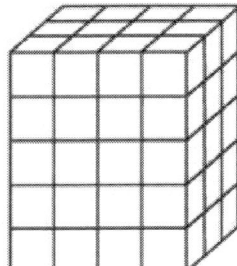

Volume of Rectangle Prisms

✎ **Find the volume of each of the rectangular prisms.**

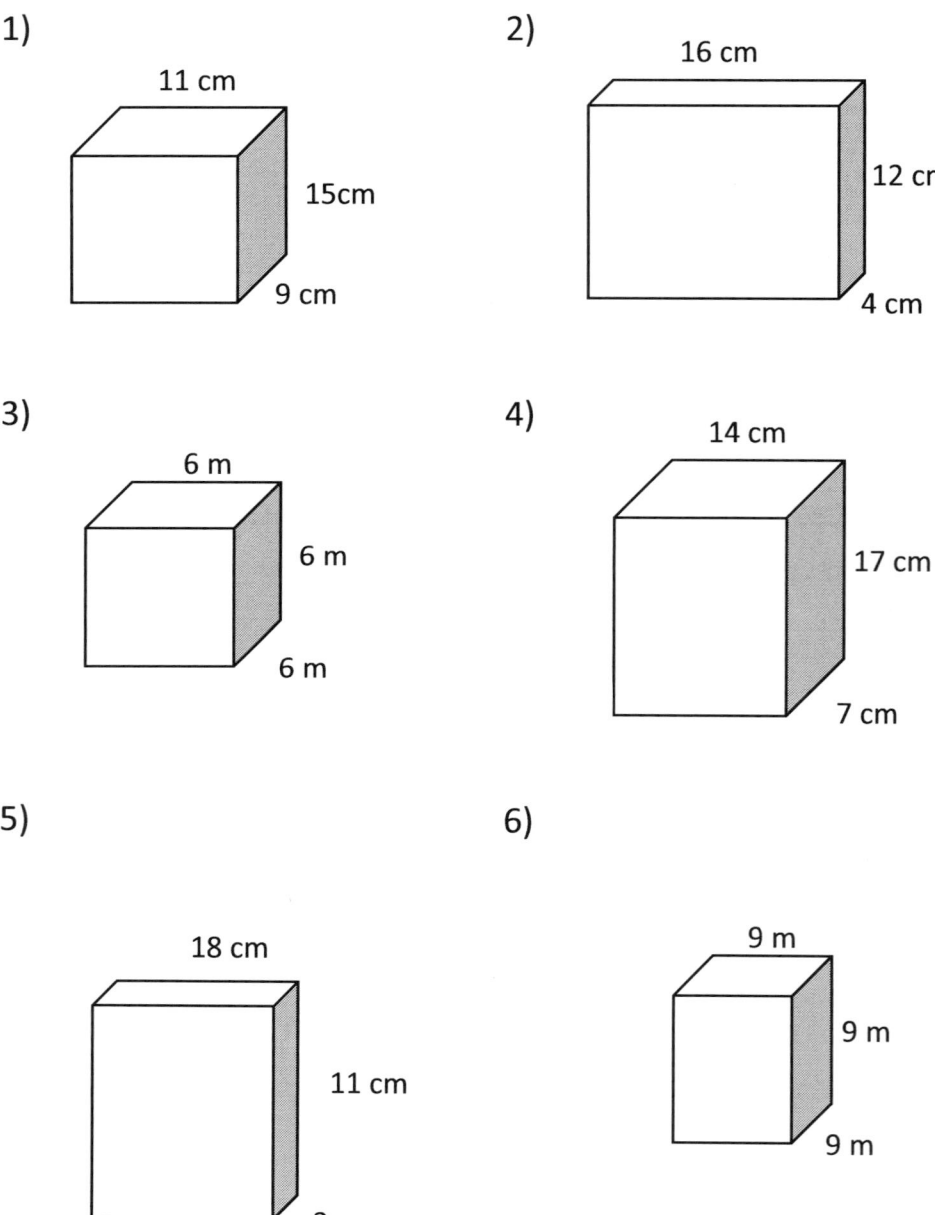

1)

11 cm

15cm

9 cm

2)

16 cm

12 cm

4 cm

3)

6 m

6 m

6 m

4)

14 cm

17 cm

7 cm

5)

18 cm

11 cm

3cm

6)

9 m

9 m

9 m

Surface Area of Cubes

✍️ **Find the surface of each cube.**

1)

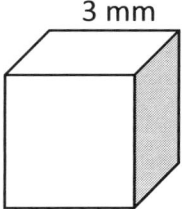

3 mm

2)

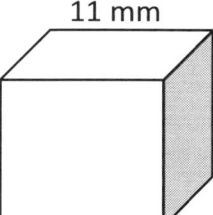

11 mm

3)

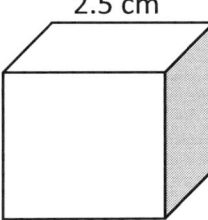

2.5 cm

4)

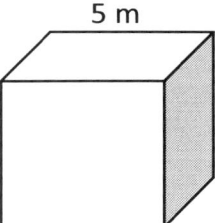

5 m

5)

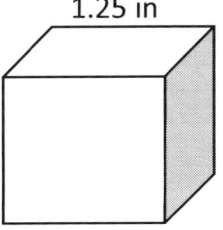

1.25 in

6)

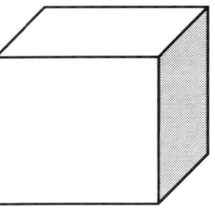

4.2 ft

Surface Area of a Rectangle Prism

✎ Find the surface of each prism.

1)

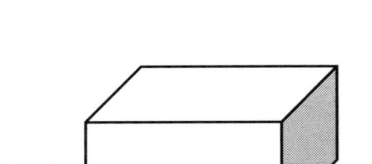

4 yd
5 yd
8 yd

2)

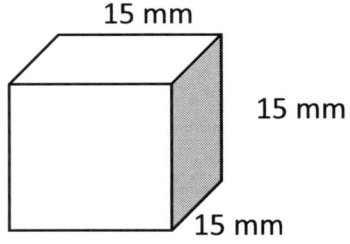

15 mm
15 mm
15 mm

3)

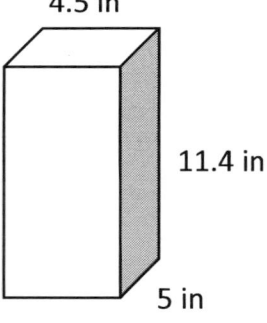

4.5 in
11.4 in
5 in

4)

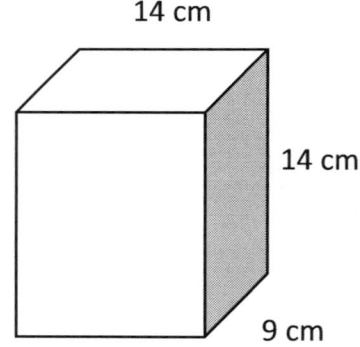

14 cm
14 cm
9 cm

Volume of a Cylinder

✎ **Find the volume of each cylinder.** ($\pi = 3.14$)

1)

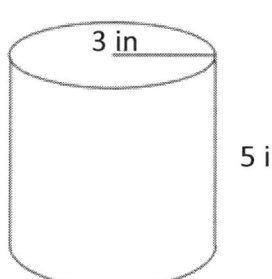

2)

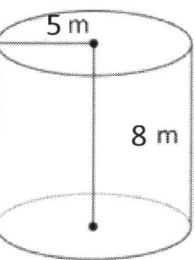

3)

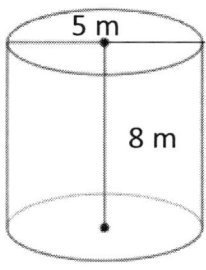

4)

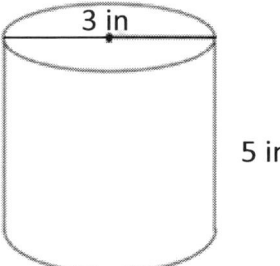

5)

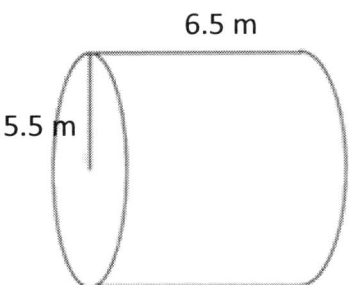

6)

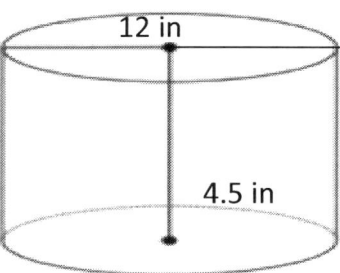

Surface Area of a Cylinder

✎ **Find the surface of each cylinder.** $(\pi = 3.14)$

1)

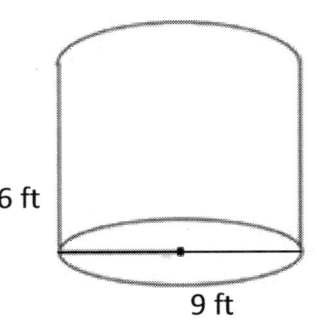

6 ft

9 ft

2)

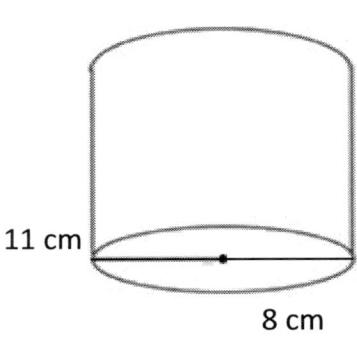

11 cm

8 cm

3)

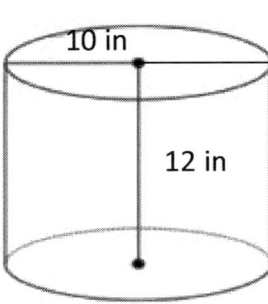

10 in

12 in

4)

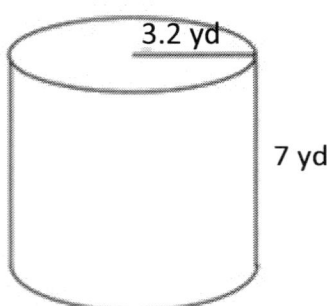

3.2 yd

7 yd

5)

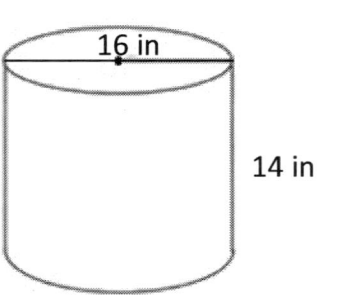

16 in

14 in

6)

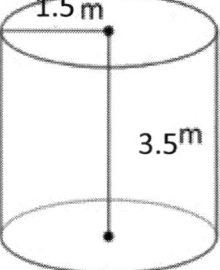

1.5 m

3.5 m

Volume of Pyramids and Cones

✎ *Find the volume of each figure*$(\pi = 3.14)$.

1)

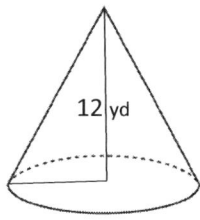

12 yd

2)

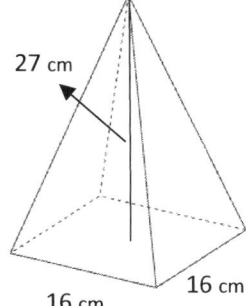

27 cm

16 cm

16 cm

3)

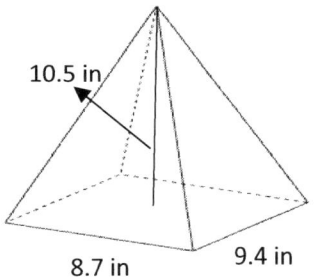

10.5 in

8.7 in

9.4 in

4)

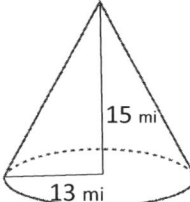

15 mi

13 mi

5)

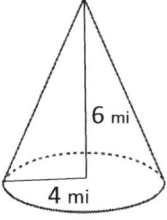

6 mi

4 mi

6)

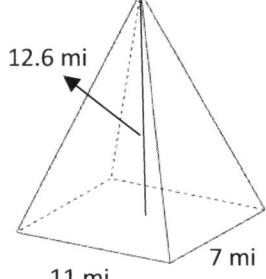

12.6 mi

11 mi

7 mi

Surface Area of Pyramids and Cones

✍ *Find the surface area of each figure*($\pi = 3.14$).

1)

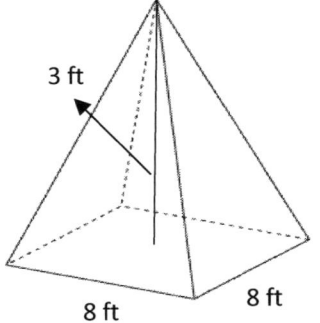

3 ft

8 ft   8 ft

2)

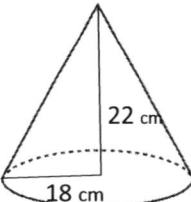

6 in

16 in   16 in

3)

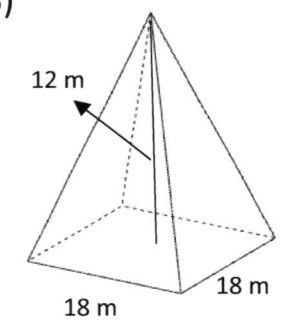

12 m

18 m

18 m

4)

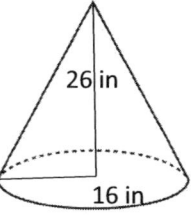

22 cm

18 cm

5)

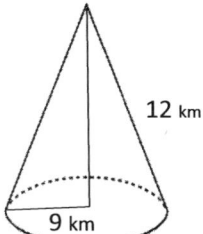

26 in

16 in

6)

12 km

9 km

Answers of Worksheets – Chapter 12

The Pythagorean Theorem

1) yes   3) yes   5) 60

2) yes   4) 40   6) 65

Angles

1) 68°   3) 52°   5) 55°   7) 45°

2) 89°   4) 17°   6) 28°   8) 72°

Area of Triangles

1) 27 mi^2 3) 108.8 m^2

2) 37.6 m^2 4) 20.13 m^2

Area of Trapezoids

1) 84 cm^2 3) 62 mi^2

2) 252 m^2 4) 55.47 nm^2

Area of Squares, Rectangles, and Parallelograms

1) A: 495 m^2, P: 91   5) A: 102 cm^2, P: 60 cm   9) P: 64 ft

2) A: 441 mm^2, P: 84   6) A: 27 mm^2, P:24 mi   10) P: 76 in

3) A: 128 ft^2, P: 56 ft   7) P: 42 m   11) P: 56 cm

4) A: 24 in^2, P: 23.2in   8) P: 48 mm   12) P: 59 ft

Area and Circumference of Circles

1) Area: 19.63 cm^2, Circumference: 15.7 cm.

2) Area: 50.24 in^2, Circumference: 25.12 in.

3) Area: 254.34 km^2, Circumference: 56.52 km.

4) Area: 94.99 m^2, Circumference: 34.54 m.

5) Area: 113.04 m^2, Circumference: 37.68 m

6) Area: 28.26 cm^2, Circumference: 18.84 cm.

7) Area: 38.47 cm², Circumference: 21.98 cm.

8) Area: 7.07 in², Circumference: 9.42 in.

Volumes of Cubes

1) 5	3) 8	5) 44
2) 36	4) 4	6) 60

Volume of Rectangle Prisms

1) 1485 cm³	3) 216 m³	5) 594 cm³
2) 768 cm³	4) 1666 cm³	6) 729 cm³

Surface Area of a Cube

1) 54 mm²	3) 37.5 cm²	5) 9.375 in²
2) 726 mm²	4) 150 m²	6) 105.84 ft²

Surface Area of a Rectangle Prism

1) 184 yd²   2) 1350 mm²   3) 261.6 in²   4) 896 cm²

Volume of a Cylinder

1) 141.3 cm³	3) 157 m³	5) 617.4 m³
2) 628 cm³	4) 35.325 m³	6) 508.68 m³

Surface Area of a Cylinder

1) 296.73 ft²	3) 533.8 in²	5) 1105.28 in²
2) 376.8 cm²	4) 204.98 yd²	6) 47.1m²

Volume of Pyramids and Cones

1) 615.4 yd³	3) 286.2 mi³	5) 100.5 in³
2) 1,024 cm³	4) 2,653.3 mi³	6) 323.4 mi³

Surface Area of Pyramids and Cones

1) $144\ in^3$	3) $864\ m^3$	5) $2,337.601\ ft^2$
2) $576\ in^2$	4) $2,623.96\ m^3$	6) 678.24 km

Chapter 13: Trigonometric Functions

Topics that you'll learn in this chapter:

- ✓ Trig ratios of General Angles
- ✓ Sketch Each Angle in Standard Position
- ✓ Finding Co–Terminal Angles and Reference Angles
- ✓ Writing Each Measure in Radians
- ✓ Writing Each Measure in Degrees
- ✓ Evaluating Each Trigonometric Expression
- ✓ Missing Sides and Angles of a Right Triangle
- ✓ Arc Length and Sector Area

Mathematics is like checkers in being suitable for the young, not too difficult, amusing, and without peril to the state. — Plato

Unlock problems

❖ Trig ratios of General Angles

θ	0°	30°	45°	60°	90°
$\sin\theta$	0	$\frac{1}{2}$	$\frac{\sqrt{2}}{2}$	$\frac{\sqrt{3}}{2}$	1
$\cos\theta$	1	$\frac{\sqrt{3}}{2}$	$\frac{\sqrt{2}}{2}$	$\frac{1}{2}$	0
$\tan\theta$	0	$\frac{\sqrt{3}}{3}$	1	$\sqrt{3}$	undefined

❖ Sketch Each Angle in Standard Position

✓ The standard position of an angle is when its vertex is located at the origin and its initial side extends along the positive x-axis.

✓ A positive angle is the angle measured in a counterclockwise direction from the initial side to the terminal side.

✓ A negative angle is the angle measured in a clockwise direction from the initial side to the terminal side.

❖ Finding Co-terminal Angles and Reference Angles

✓ Co-terminal angles are equal angles.

✓ To find a co-terminal of an angle, add or subtract 360 degrees (or 2π for radians) to the given angle.

✓ Reference angle is the smallest angle that you can make from the terminal side of an angle with the x-axis.

❖ Writing Each Measure in Radians

✓ Radians = degrees$\times \frac{\pi}{180}$

Example: Convert 150 degrees to radians.

$$\text{Radians} = 150 \times \frac{\pi}{180} = \frac{5\pi}{6}$$

❖ Writing Each Measure in Degrees

✓ Degrees = radians$\times \frac{180}{\pi}$

Example: Convert $\frac{2\pi}{3}$ to degrees.

$$\frac{2\pi}{3} \times \frac{180}{\pi} = \frac{360\pi}{3\pi} = 120$$

❖ Evaluating Each Trigonometric Function

✓ Step 1: Draw the terminal side of the angle.

✓ Step 2: Find reference angle. (It is the smallest angle that you can make from the terminal side of an angle with the x-axis.)

✓ Step 3: Find the trigonometric function of the reference angle.

❖ Missing Sides and Angles of a Right Triangle

✓ SOH − CAH - TOA

✓ sine $\theta = \frac{\text{opposite}}{\text{hypotenuse}}$ (SOH)

✓ Cos $\theta = \frac{\text{adjacent}}{\text{hypotenuse}}$ (CAH)

✓ $\tan \theta = \frac{\text{opposite}}{\text{adjacent}}$ (TOA)

❖ Arc Length and Sector Area

✓ Area of a sector $= \frac{1}{2}r^2\theta$

✓ length of a sector $= (\frac{\theta}{180})\pi r$

Trig ratios of General Angles

✎ *Use a calculator to find each. Round your answers to the nearest ten–thousandth.*

1) $\sin - 150°$

2) $\sin 120°$

3) $\cos 315°$

4) $\cos 120°$

5) $\sin 180°$

6) $\sin - 300°$

✎ *Find the exact value of each trigonometric function. Some may be undefined.*

7) $\sec -120$

8) $\tan - \dfrac{3\pi}{4}$

11) $\cos \dfrac{\pi}{6}$

12) $\cot \dfrac{\pi}{3}$

13) $\sec - \dfrac{3\pi}{2}$

14) $\tan - \dfrac{2\pi}{3}$

Sketch Each Angle in Standard Position

*Draw the angle with the given measure in standard position.*

1) $140°$

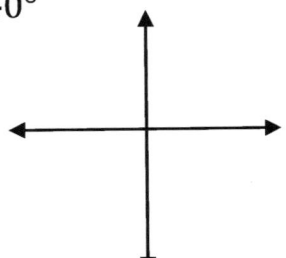

4) $\dfrac{55\pi}{12}$

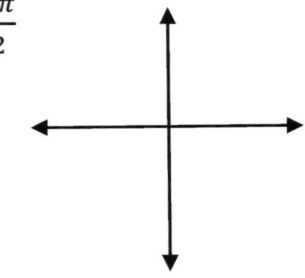

2) $-250°$

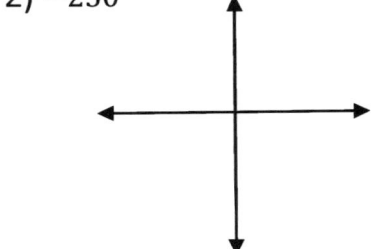

5) $\dfrac{5\pi}{6}$

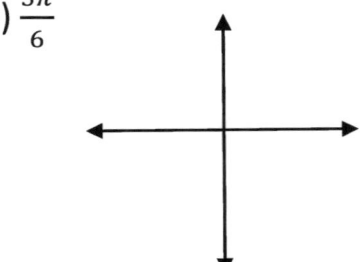

3) $610°$

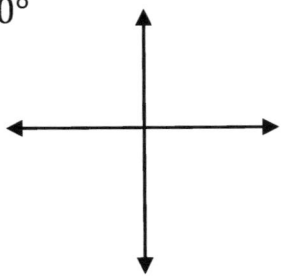

6) $-\dfrac{13\pi}{6}$

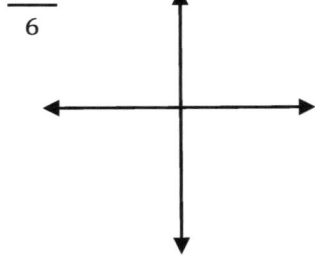

Finding Co-terminal Angles and Reference Angles

✍ *Find a conterminal angle between 0° and 360°.*

1) $-480°$

2) $680°$

3) $-335°$

4) $-430°$

✍ *Find a coterminal angle between 0 and 2π for each given angle.*

5) $\dfrac{11\pi}{3}$

6) $-\dfrac{5\pi}{6}$

7) $-\dfrac{2\pi}{45}$

8) $\dfrac{14\pi}{3}$

✍ *Find the reference angle.*

9)

10)

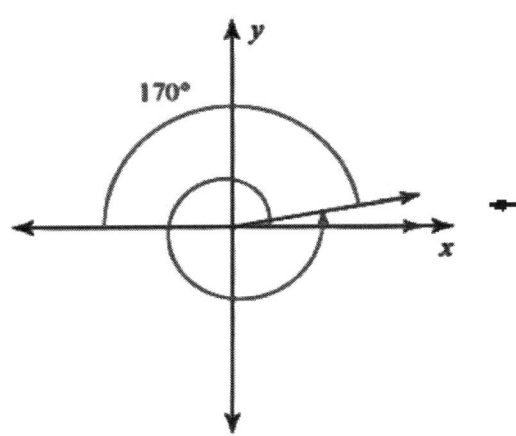

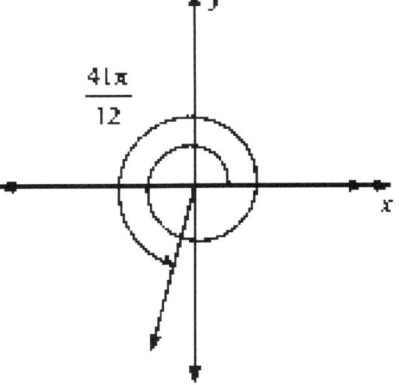

Writing Each Measure in Radians

✍️ *Convert each degree measure into radians.*

1) −120°

2) 220°

3) 160°

4) 920°

5) −200°

6) 230°

7) 265°

8) 20°

9) 420°

10) 30°

11) 297°

12) 500°

13) 504°

14) −130°

15) −260°

16) 423°

17) 440°

18) −190°

19) 250°

20) 350°

Writing Each Measure in Degrees

✍️*Convert each radian measure into degrees.*

1) $\dfrac{\pi}{60}$

2) $\dfrac{12\pi}{30}$

3) $\dfrac{13\pi}{6}$

4) $\dfrac{\pi}{3}$

5) $-\dfrac{10\pi}{9}$

6) $\dfrac{12\pi}{3}$

7) $-\dfrac{16\pi}{3}$

8) $-\dfrac{8\pi}{20}$

9) $\dfrac{5\pi}{6}$

10) $\dfrac{2\pi}{9}$

11) $\dfrac{7\pi}{6}$

12) $\dfrac{15\pi}{60}$

13) $\dfrac{11\pi}{4}$

14) $-\dfrac{22\pi}{11}$

15) $\dfrac{14\pi}{9}$

16) $-\dfrac{41\pi}{60}$

17) $-\dfrac{17\pi}{6}$

18) $\dfrac{32\pi}{18}$

19) $-\dfrac{4\pi}{3}$

20) $\dfrac{5\pi}{9}$

Evaluating Each Trigonometric Function

📝 *Find the exact value of each trigonometric function.*

1) $\cos 315°$

2) $\cot \dfrac{\pi}{6}$

3) $\tan \dfrac{\pi}{6}$

4) $\cot -\dfrac{5\pi}{6}$

5) $\cos \dfrac{2\pi}{3}$

6) $\cos - 240°$

7) $\sin 480°$

8) $\tan 480°$

9) $\cot 390°$

10) $\tan 405°$

📝 Use the given point on the terminal side of angle θ to find the value of the trigonometric function indicated.

11) $\sin θ; (-6, 4)$

12) $\cos θ; (2, -2)$

13) $\cot θ; (-7, \sqrt{15})$

14) $\cos θ; (-2\sqrt{3}, -2)$

15) $\sin θ; (-\sqrt{7}, 3)$

16) $\tan θ; (-11, -2)$

Missing Sides and Angles of a Right Triangle

✎ *Find the value of each trigonometric ratio as fractions in*

their simplest form.

1) tan A

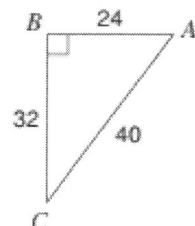

2) sin X

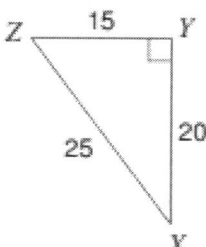

✎ *Find the missing side. Round answers to the nearest*

tenth.

3)

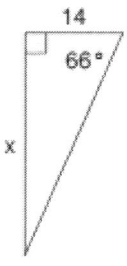

4)

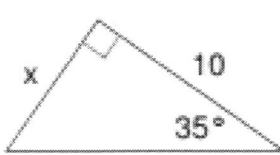

5)

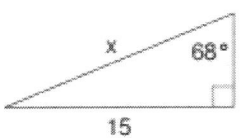

6)

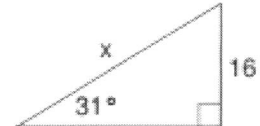

Arc Length and Sector Area

✍ *Find the length of each arc. Round your answers to the nearest tenth.*

1) r = 12 cm, θ = 65°   3) r = 33 ft, θ = 90°

2) r = 10 ft, θ = 95°   4) r = 16 m, θ = 86°

✍ *Find area of a sector. Do not round.*

5)   7)

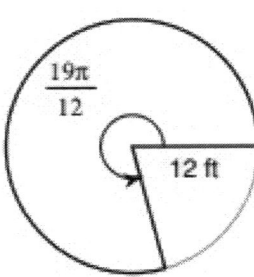

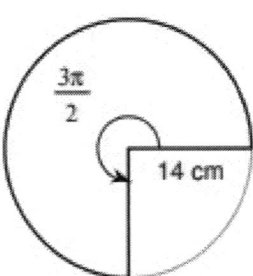

6)   8)

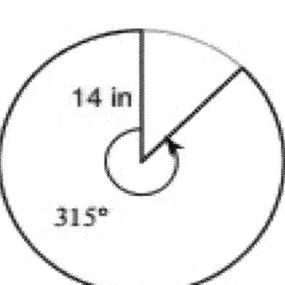

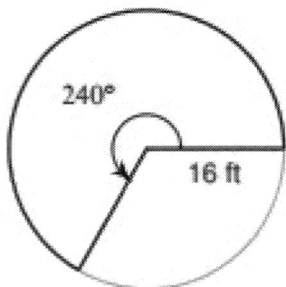

Answers of Worksheets – Chapter 13

Trig Ratios of General Angles

1) $-\dfrac{1}{2}$

2) $\dfrac{\sqrt{3}}{2}$

3) $\dfrac{\sqrt{2}}{2}$

4) $-\dfrac{1}{2}$

5) 0

6) $\dfrac{\sqrt{3}}{2}$

7) -2

8) 1

9) $\dfrac{\sqrt{3}}{2}$

10) $\dfrac{\sqrt{3}}{3}$

11) Undefined

12) $\sqrt{3}$

Sketch Each Angle in Standard Position

1) 140

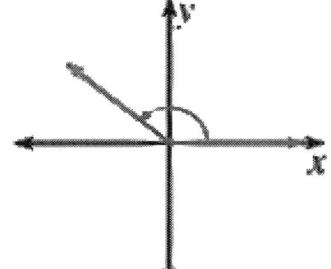

4) $\dfrac{55\pi}{12}$

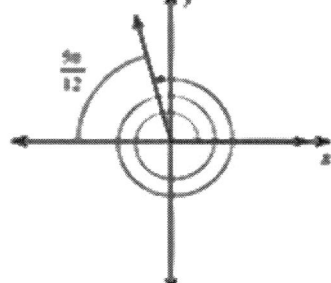

2) -250

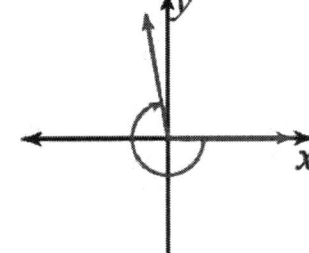

5) $\dfrac{5\pi}{6}$

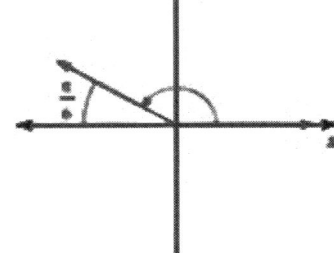

3) 610

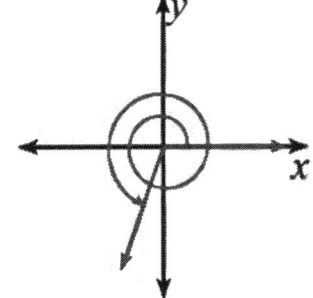

6) $\dfrac{11\pi}{6}$

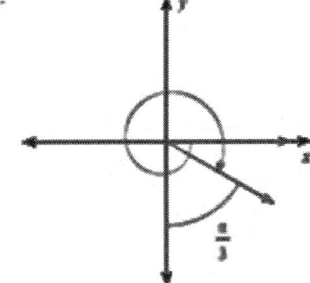

Finding Co–Terminal Angles and Reference Angles

1) $240°$

2) $320°$

3) $25°$

4) $290°$

5) $\dfrac{5\pi}{3}$

6) $\dfrac{7\pi}{6}$

7) $\dfrac{88\pi}{45}$

8) $\dfrac{2\pi}{3}$

9) $370°$

10) $\dfrac{5\pi}{12}$

Writing Each Measure in Radians

1) $-\dfrac{2\pi}{3}$

2) $\dfrac{11\pi}{9}$

3) $\dfrac{8\pi}{9}$

4) $\dfrac{46\pi}{9}$

5) $-\dfrac{10\pi}{9}$

6) $\dfrac{23\pi}{18}$

7) $\dfrac{53\pi}{36}$

8) $\dfrac{\pi}{9}$

9) $\dfrac{7\pi}{3}$

10) $\dfrac{\pi}{6}$

11) $\dfrac{33\pi}{2}$

12) $\dfrac{25\pi}{9}$

13) $\dfrac{14\pi}{5}$

14) $-\dfrac{13\pi}{18}$

15) $-\dfrac{13\pi}{9}$

16) $\dfrac{47\pi}{2}$

17) $\dfrac{22\pi}{9}$

18) $-\dfrac{19\pi}{18}$

19) $\dfrac{25\pi}{18}$

20) $\dfrac{35\pi}{18}$

Writing Each Measure in Degrees

1) $3°$

2) $72°$

3) $390°$

4) $60°$

5) $-200°$

6) $720°$

7) $-960°$

8) $-72°$

9) $150°$

10) $40°$

11) $210°$

12) $45°$

13) $495°$

14) $-360°$

15) $280°$

16) $-123°$

17) $-510°$

18) $320°$

19) $-240°$

20) $100°$

Evaluating Each Trigonometric Expression

1) $\dfrac{\sqrt{2}}{2}$

2) $\sqrt{3}$

3) $-\sqrt{3}$

4) $\sqrt{3}$

5) $-\dfrac{1}{2}$

6) $-\dfrac{1}{2}$

7) $\dfrac{\sqrt{3}}{2}$

8) $-\sqrt{3}$

9) $\sqrt{3}$

10) 1

11) $\dfrac{2\sqrt{13}}{13}$

12) $-\sqrt{2}$

13) $-\dfrac{7\sqrt{15}}{15}$

14) $-\dfrac{\sqrt{3}}{2}$

15) $\dfrac{3}{4}$

16) $\dfrac{2}{11}$

Missing Sides and Angles of a Right Triangle

1) $\frac{4}{3}$

2) $\frac{3}{5}$

3) 31.4

4) 7.0

5) 16.2

6) 31.1

Arc Length and Sector Area

1) 74 cm

2) 17 ft

3) 52 ft

4) 24 m

5) 114π ft^2

6) $\frac{343\pi}{2}$ in^2

7) 147π cm^2

8) $\frac{512\pi}{3}$ ft^2

Chapter 14: Statistics

Topics that you'll learn in this chapter:

✓ Mean, Median, Mode, and Range of the Given Data

✓ Box and Whisker Plots

✓ Bar Graph

✓ Stem– And– Leaf Plot

✓ The Pie Graph or Circle Graph

✓ Dot and Scatter Plots

✓ Probability of Simple Events

Mathematics is no more computation than typing is literature.

— John Allen Paulos

Unlock problems

❖ Mean, Median, Mode, and Range of the Given Data

✓ Mean: $\frac{\text{sum of the data}}{\text{of data entires}}$

✓ Median: Middle value in the list of number

✓ Mode: value in the list that appears most often

✓ Range: largest value − smallest value

Example:

22, 16, 12, 9, 7, 6, 4, 6

Mean = 10.25

Median: 8. first order the numbers (4, 6, 6, 7, 9, 12, 16, 22)

If number even: the average of two middle ($\frac{7+9}{2} = \frac{16}{2} = 8$)

Mod = 6

Range = 18

❖ Box and Whisker Plots

✓ Box–and–whisker plots display data including quartiles.

✓ IQR − interquartile range shows the difference from Q1 to Q3.

✓ Extreme Values are the smallest and largest values in a data set.

Example:

73, 84, 86, 95, 68, 67, 100, 94, 77, 80, 62, 79

Maximum: 100, Minimum: 62; Q_1: 70.5; Q_2: 79.5; Q_3: 90

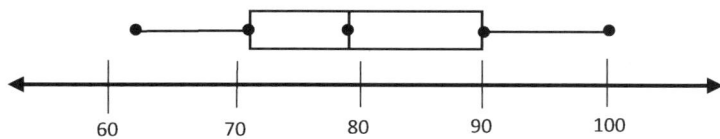

❖ Bar Graph

✓ A bar graph is a chart that presents data with bars in different heights to match with the values of the data. The bars can be graphed horizontally or vertically.

❖ Stem–And–Leaf Plot

✓ Stem–and–leaf plots display the frequency of the values in a data set.

✓ We can make a frequency distribution table for the values, or we can use a stem–and–leaf plot.

Example:

56, 58, 42, 48, 66, 64, 53, 69, 45, 72

Stem	leaf		
4	2	5	8
5	3	6	8
6	4	6	9
7	2		

❖ The Pie Graph or Circle Graph

✓ A Pie Chart is a circle chart divided into sectors; each sector represents the relative size of each value.

❖ Scatter Plots

✓ A Scatter (xy) Plot shows the values with points that represent the relationship between two sets of data.

✓ The horizontal values are usually x and vertical data is y.

❖ Probability of Simple Events

✓ Probability is the likelihood of something happening in the future. It is expressed as a number between zero (can never happen) to 1 (will always happen).

✓ Probability can be expressed as a fraction, a decimal, or a percent.

Example:

Probability of a flipped coins turns up 'heads' is $0.5 = \frac{1}{2}$

❖ Experimental Probability

✓ Experimental probability refers to the probability of an event occurring when an experiment was conducted.

✓ Experimental probability $= \frac{\text{Number of event occurrences}}{\text{Total number of trials}}$

❖ Factorials

✓ Factorials means to multiply a series of descending natural numbers.

Example:

5!$= 5 \times 4 \times 3 \times 2 \times 1$

$5! = 120$

❖ Permutations

✓ The number of ways to choose a sample of r elements from a set of n distinct objects where order does matter, and replacements are not allowed.

$$_nP_k = P_k^n = \frac{n!}{(n-k)!}$$

Example:

$$_4P_2 = \frac{4!}{(4-2)!} = 12$$

❖ Combination

✓ The number of ways to choose a sample of r elements from a set of n distinct objects where order does not matter, and replacements are not allowed.

$$_nC_r = C_r^n = \frac{n!}{r! \ (n-r)!}$$

Example:

$$_4C_2 = \frac{4!}{2!(4-2)!} = 3$$

Mean and Median

🖎 *Find Mean and Median of the Given Data.*

1) $8, 12, 5, 3, 2$

2) $3, 6, 3, 7, 4, 13$

3) $13, 5, 1, 7, 9$

4) $6, 4, 2, 7, 3, 2$

5) $6, 5, 7, 5, 7, 1, 11$

6) $6, 1, 4, 4, 9, 2, 19$

7) $12, 4, 1, 5, 9, 7, 7, 19$

8) $18, 9, 5, 4, 9, 6, 12$

9) $28, 25, 15, 16, 32, 44, 71$

10) $10, 5, 1, 5, 4, 5, 8, 10$

11) $18, 15, 30, 64, 42, 11$

12) $44, 33, 56, 78, 41, 84$

🖎 *Solve.*

13) In a javelin throw competition, five athletics score 56, 58, 63, 57 and 61 meters. What are their Mean and Median?

_____

14) Eva went to shop and bought 3 apples, 5 peaches, 8 bananas, 1 pineapple and 3 melons. What are the Mean and Median of her purchase? _____

Mode and Range

✎ *Find Mode and Rage of the Given Data.*

1) 8, 2, 5, 9, 1, 2

Mode: _____ Range: _____

2) 6, 6, 2, 3, 6, 3, 9, 12

Mode: _____ Range: _____

3) 4, 4, 3, 9, 7, 9, 4, 6, 4

Mode: _____ Range: _____

4) 12, 9, 2, 9, 3, 2, 9, 5

Mode: _____ Range: _____

5) 9, 5, 9, 5, 8, 9, 8

Mode: _____ Range: _____

6) 0, 1, 4, 10, 9, 2, 9, 1, 5, 1

Mode: _____ Range: _____

7) 6, 5, 6, 9, 7, 7, 5, 4, 3, 5

Mode: _____ Range: _____

8) 7, 5, 4, 9, 6, 7, 7, 5, 2

Mode: _____ Range: _____

9) 2, 2, 5, 6, 2, 4, 7, 6, 4, 9

Mode: _____ Range: _____

10) 7, 5, 2, 5, 4, 5, 8, 10

Mode: _____ Range: _____

11) 4, 1, 5, 2, 2, 12, 18, 2

Mode: _____ Range: _____

12) 6, 3, 5, 9, 6, 6, 3, 12

Mode: _____ Range: _____

✎ *Solve.*

13) A stationery sold 12 pencils, 36 red pens, 44 blue pens, 12 notebooks, 18 erasers, 34 rulers and 32 color pencils. What are the Mode and Range for the stationery sells?

Mode: _____   Range: _____

14) In an English test, eight students score 14, 13, 17, 11, 19, 20, 14 and 15. What are their Mode and Range?

_____

Times Series

✍ *Use the following Graph to complete the table.*

Day	Distance (km)
1	
2	

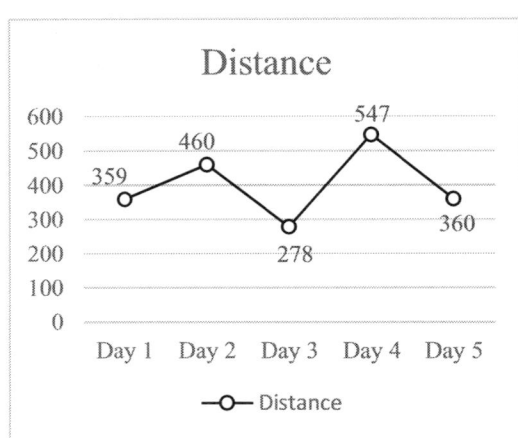

The following table shows the number of births in the US from 2007 to 2012 (in millions).

Year	Number of births (in millions)
2007	4.32
2008	4.25
2009	4.13
2010	4
2011	3.95
2012	3.95

Draw a times series for the table.

Box and Whisker Plot

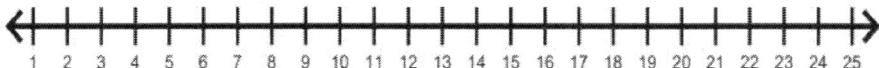 *Make box and whisker plots for the given data.*

$$1, 5, 18, 8, 3, 11, 13, 12, 24, 17, 10, 15, 25$$

Bar Graph

Graph the given information as a bar graph.

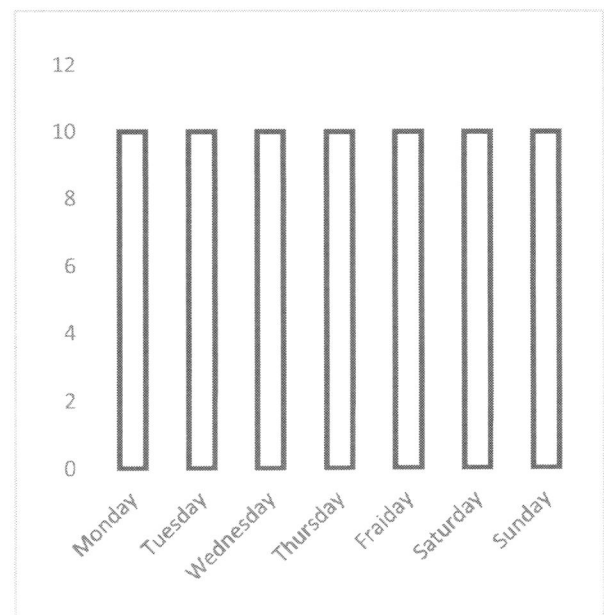

Day	Sale House
Monday	5
Tuesday	7
Wednesday	9
Thursday	8
Friday	0
Saturday	3
Sunday	4

Dot plots

A survey of "How many pets each person owned?" has these results:

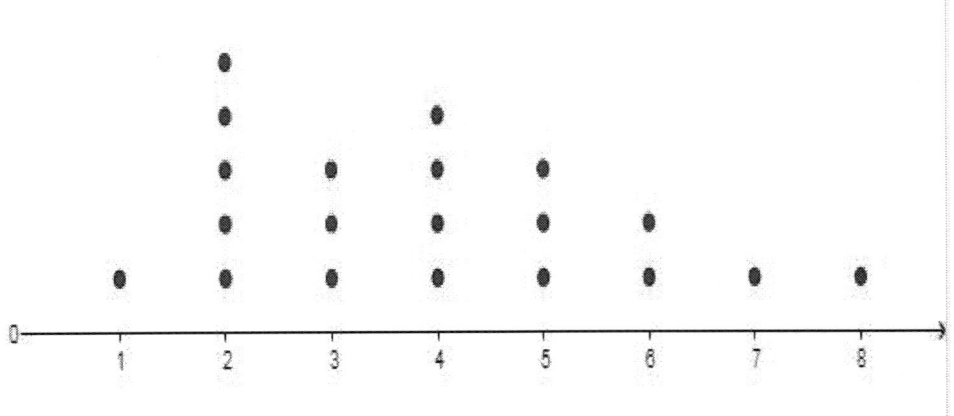

3) What is the most common number of pets?

4) How many people have 3 or less than 3 pets?

5) How many people have more than 6 pets?

Scatter Plots

✎ *Construct a scatter plot.*

x	1	2	3	4	4.5	5
y	5	3.5	4	2	7	1.5

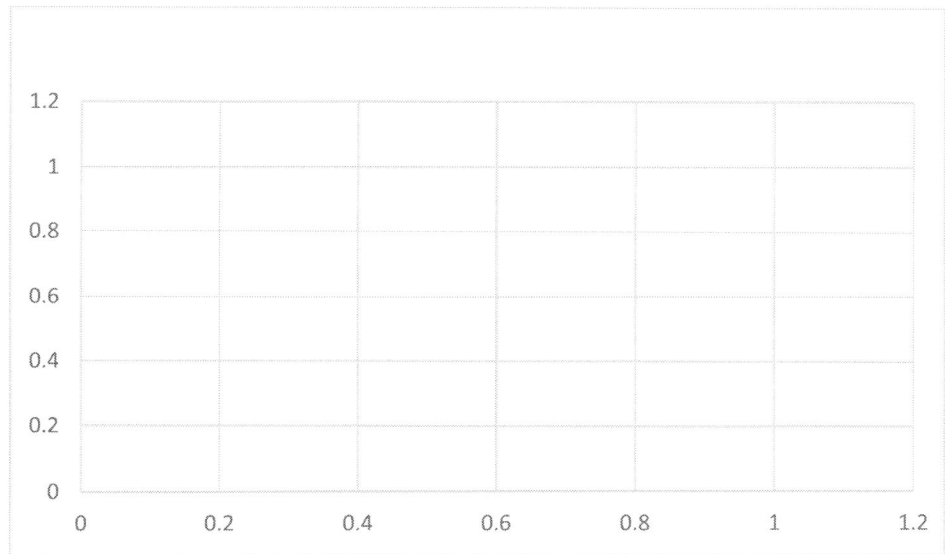

Stem–And–Leaf Plot

✍ *Make stem ad leaf plots for the given data.*

1) 22, 24, 27, 21, 52, 24, 58, 57, 29, 24, 19, 12

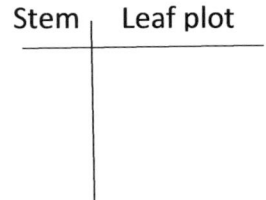

2) 11, 45, 34, 18, 15, 11, 32, 41, 40, 30, 45, 35

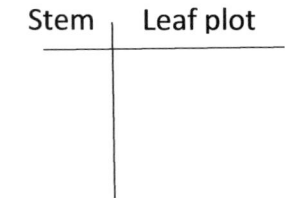

3) 112, 87, 96, 85, 100, 117, 92, 114, 88, 112, 98, 107

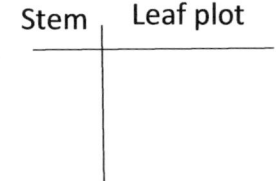

4) 63, 50, 104, 63, 72, 56, 109 63, 75, 59, 63, 108, 79

The Pie Graph or Circle Graph

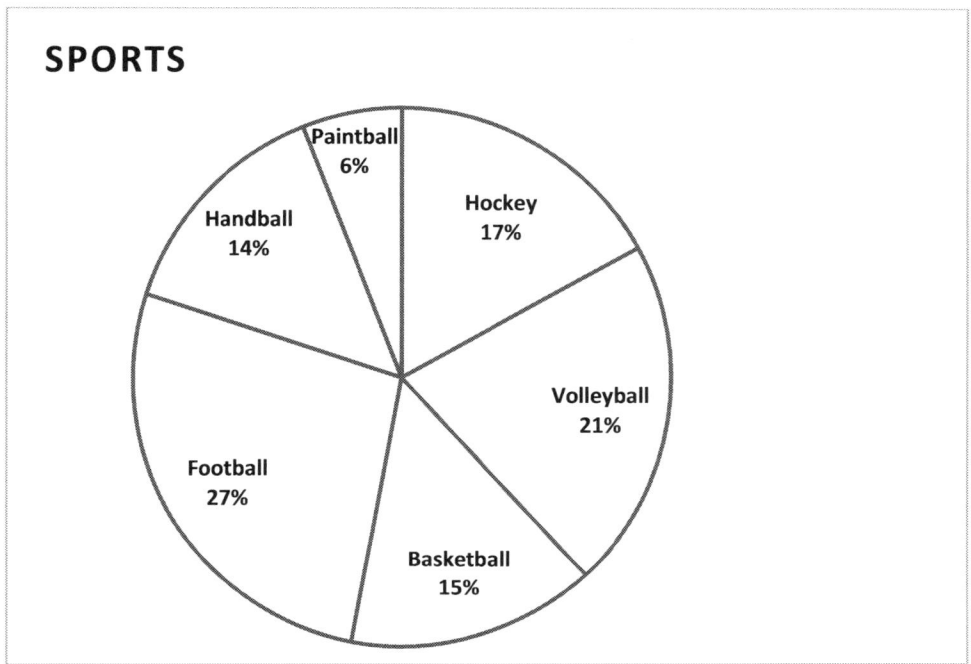

SPORTS

Favorite Sports:

1) What percentage of pie graph is Handball?

2) What percentage of pie graph is Hockey?

3) Which sport is the most?

4) Which sport is the least?

5) What percentage of pie graph is Volleyball?

6) What percentage of pie not football and volleyball?

Probability of Simple Events

✍ *Solve.*

1) A number is chosen at random from 25 to 34. Find the probability of selecting an even number.

2) A number is chosen at random from 21 to 60. Find the probability of selecting multiples of 5.

3) Find the probability of selecting 4 aces from a deck of card.

4) A number is chosen at random from 10 to 19. Find the probability of selecting of 11 and factors of 3.

5) What probability of selecting a ball less than 11 from 40 different bingo balls?

6) Find the probability of not selecting a king from a deck of card.

Experimental Probability

On cube	frequency
1	6
2	9
3	5
4	7
5	8
6	5

1) Theoretically if you roll a number cube 24 times, how many times would you expect to roll the number two?

2) How many times did you roll the number two in the experiment?

3) Is there any difference between theoretical and experimental probability?

4) What is the theoretical probability for rolling a number greater than 4?

5) What was the experimental probability of rolling a number greater than 3?

Factorials

Determine the value for each expression.

1) $5!$

2) $\dfrac{7!}{10!}$

3) $\dfrac{10!}{6!}$

4) $\dfrac{n!}{(n-3)!}$

5) $\dfrac{12!}{8!4!}$

6) $\dfrac{45!}{44!}$

7) $\dfrac{100!}{101!}$

8) $\dfrac{(M+1)!}{(M-1)!}$

9) $\dfrac{15!}{10!}$

10) $\dfrac{28!}{25!}$

11) $\dfrac{0!4!}{1!0!}$

12) $\dfrac{22!}{20!}$

13) $7! - 3!$

14) $7! + 3!$

Permutations

✍ *Evaluate each expression.*

1) $5 \; _4P_1$

2) $_7P_3$

3) $_8P_5$

4) $12 + \; _{10}P_3$

5) $P(5,3)$

6) $P(6,2)$

7) $(\; _5P_4)$

8) $\frac{1}{2}(\; _{12}P_1)$

9) $_4P_0$

10) $_0P_0$

11) $_4P_4$

12) $_9P_3$

13) How many possible 7–digit telephone numbers are there? Someone left their umbrella on the subway and we need to track them down.

14) With repetition allowed, how many ways can one choose 6 out of 15 things?

Combination

✏️ *List all possible combinations.*

1) 1, 4, 3, 5, taken four at a time

2) A, B, D, taken two at a time

✏️ *Evaluate each expression.*

3) $_4C_1$

9) $5(\,_{13}C_9)$

4) $_7C_3$

10) $\frac{1}{2}(\,_{12}C_1)$

5) $\binom{12}{5}$

11) $_5C_0$

6) $3 + \binom{21}{14}$

12) $_0C_0$

7) $C(5,3)$

13) $_6C_6$

8) $C(6,2)$

14) $_{18}C_{17}$

Answers of Worksheets – Chapter 14

Mean and Median

1) Mean: 6, Median: 5

2) Mean: 6, Median: 5

3) Mean: 7, Median: 7

4) Mean: 4, Median: 3.5

5) Mean: 6, Median: 6

6) Mean: 8, Median: 4

7) Mean: 8, Median: 7

8) Mean: 9, Median: 9

9) Mean: 33, Median: 28

10) Mean: 6, Median: 5

11) Mean: 30, Median: 24

12) Mean: 56, Median: 50

13) Mean: 59, Median: 58

14) Mean: 4, Median: 3

Mode and Range

1) Mode: 2, Range: 8

2) Mode: 6, Range: 10

3) Mode: 4, Range: 6

4) Mode: 9, Range: 10

5) Mode: 9, Range: 4

6) Mode: 1, Range: 10

7) Mode: 5, Range: 6

8) Mode: 7, Range: 7

9) Mode: 2, Range: 7

10) Mode: 5, Range: 8

11) Mode: 2, Range: 17

12) Mode: 6, Range: 9

13) Mode: 12, Range: 32

14) Mode: 14, Range: 9

Times Series

Day	Distance (km)
1	359
2	460
3	278
4	547
5	360

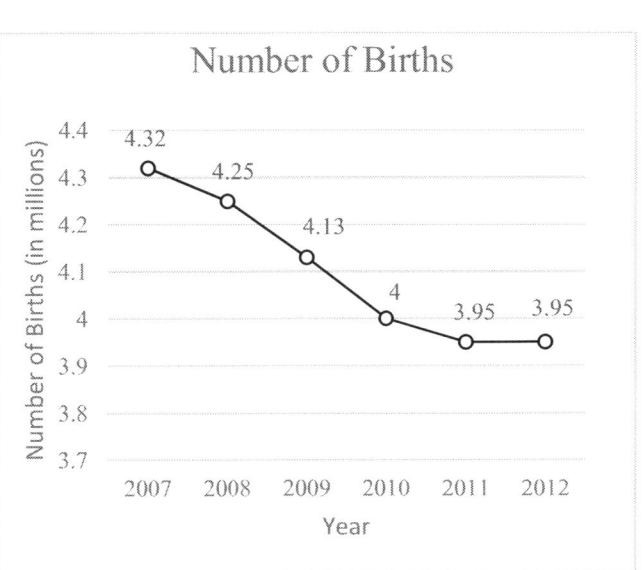

Box and Whisker Plots

1, 3, 5, 8, 10, 11, 12, 13, 15, 17, 18, 24, 25

Maximum: 25, Minimum: 1, Q_1: 6.5, Q_2: 12, Q_3: 17.5

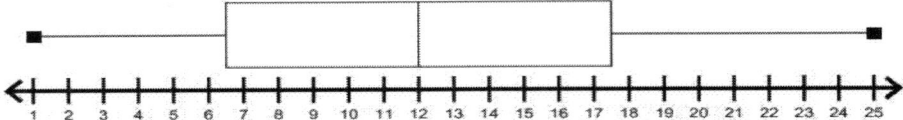

Dot plots

1) *11*   3) *2*   5) *2*

2) *5*   4) *9*

Bar Graph

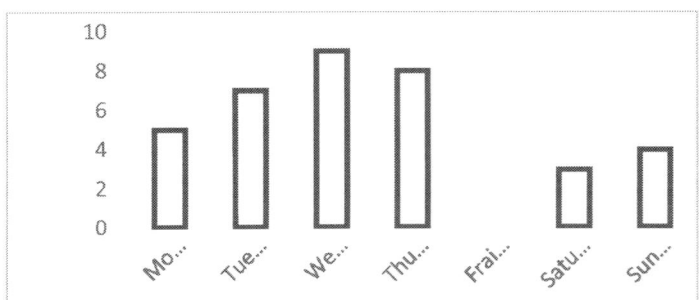

Scatter Plots

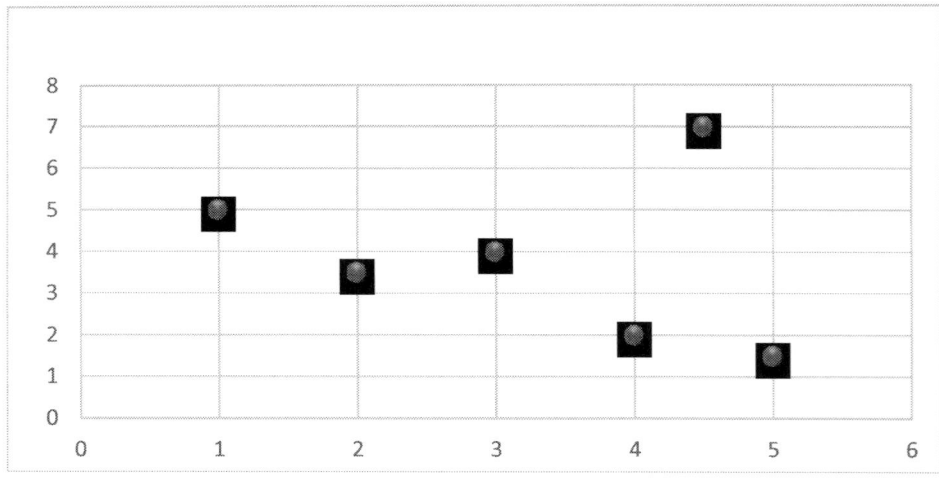

Stem–And–Leaf Plot

1)

Stem	leaf
1	2 9
2	1 2 4 4 4 7 9
5	2 7 8

2)

Stem	leaf
1	1 1 5 8
3	0 2 4 5
4	0 1 5 5

3)

Stem	leaf
8	5 7 8
9	2 6 8
10	0 7
11	2 2 4 7

4)

Stem	leaf
5	0 6 9
6	3 3 3 3
7	2 5 9
10	4 8 9

The Pie Graph or Circle Graph

1) 14%

2) Football

3) Paintball

4) 17%

5) 21%

6) 52%

Probability of simple events

1) $\frac{1}{2}$

2) $\frac{1}{5}$

3) $\frac{1}{13}$

4) $\frac{2}{5}$

5) $\frac{1}{4}$

6) $\frac{12}{13}$

Experimental Probability

1) 4

2) 9

3) yes

4) 1/3

5) 1/2

Factorials

1) 120

2) $\frac{1}{720}$

3) 5,040

4) $n(n-1)(n-2)$

5) 495

6) 45

7) $\frac{1}{101}$

8) $M(M-1)$

9) 360,360

10) 19,656

11) 24

12) 462

13) 5034

14) 5046

Permutations

1) 20

2) 210

3) 6720

4) 732

5) 60

6) 30

7) 120

8) 6

9) 1

10) 1

11) 24

12) 504

13) 10^7

14) 15^6

Combination

1) 1435

2) AB, AD, BD

3) 4

4) 35

5) 792

6) 116,283

7) 10

8) 15

9) 3,575

10) 6

11) 1

12) 1

13) 1

14) 18

ALEKS Math Test Review

ALEKS (Assessment and Learning in Knowledge Spaces) is an artificial intelligence-based assessment tool to measure students' mathematical knowledge to place students in the most appropriate level for their current math skills.

ALEKS test consists 20 to 35 open-ended questions covering a variety of math topics and there's no time limit on the test, so you can focus on doing your best to demonstrate your skills.

ALEKS uses a computer–adaptive technology and the questions you see are based on your skill level. Your response to each question drives the difficulty level of the next question.

ALEKS does NOT permit the use of personal calculators on the placement test. The test expects students to be able to answer certain questions without the assistance of a calculator. Therefore, they provide an onscreen calculator for students to use on some questions.

In this section, there are two complete ALEKS Mathematics Placement Assessment Tests.

Take these tests to see what score you'll be able to receive on a real ALEKS test.

The hardest arithmetic to master is that which enables us to count our blessings. ~Eric Hoffer

Time to Test

Time to refine your skill with a practice examination

Take a practice ALEKS Math Test to simulate the test day experience. After you've finished, score your test using the answer key.

Before You Start

- You'll need a pencil, a calculator and a timer to take the test.
- After you've finished the test, review the answer key to see where you went wrong.

Good Luck!

ALEKS Math Practice Test Answer Sheets

Remove (or photocopy) these answer sheets and use them to complete the practice tests.

ALEKS Practice Test

1		19	
2		20	
3		21	
4		22	
5		23	
6		24	
7		25	
8		26	
9		27	
10		28	
11		29	
12		30	
13		31	
14		32	
15		33	
16		34	
17		35	
18			

ALEKS Math Practice Test 1

Mathematics Placement Assessment

❖ **35 Questions.**

❖ **Total time for this test: No Limit Time.**

❖ **You may NOT use a calculator on this Section.**

Administered *Month Year*

1) If x is a positive integer divisible by 4, and $x < 40$, what is the greatest possible value of x?

2) If $a = 5$, what is the value of b in this equation?

$$b = \frac{a^2}{5} + 5$$

3) Two third of 15 is equal to $\frac{2}{5}$ of what number?

4) $(p^3).\ (p^6) =$ _____

5) Simplify: $8 - \frac{2}{5}x \geq 10$

6) A soccer team played 150 games and won 60 percent of them. How many games did the team win?

7) What is a common factor of both $x^2 - x - 6$ and $x^2 - 5x + 6$?

8) The ratio of boys to girls in a school is 5:3. If there are 400 students in a school, how many boys are in the school.

9) What is the area of an isosceles right triangle that has one leg that measures 8 cm?

10) $(x + 3)(x + 5) =$

11) $\frac{1}{5b^2} + \frac{1}{6b} = \frac{1}{b^2}$, then b =?

12) If two angles in a triangle measure 55 degrees and 42 degrees, what is the value of the third angle?

13) $4^{\frac{7}{3}} \times 4^{\frac{2}{3}}$ =?

14) What is 163.7599 rounded to the nearest hundredth?

15) $x^2 - 81 = 0$, What is (are) the value of x?

16) If $a = 2$ what's the value of $3a^2 + 4a + 12$?

17) Sophia purchased a sofa for $469.50. The sofa is regularly priced at $626. What was the percent discount Sophia received on the sofa?

18) The average of five consecutive numbers is 34. What is the smallest number?

19) The sum of three numbers is 65. If another number is added to these three numbers, the average of the four numbers is 30. What is the fourth number?

20) If $f(x) = 4 + x$ and $g(x) = -x^2 - 2 - 3x$, then find $(g - f)(x)$?

21) David owed $7360. After making 42 payments of $125 each,

how much did he have left to pay?

22) Simplify the following inequality.

$$\frac{|3+x|}{8} \leq 3$$

23) $\tan\left(-\frac{\pi}{4}\right) =?$

24) $\frac{\sqrt{27a^5b^3}}{\sqrt{3ab^2}} = ?$

25) Find the slope–intercept form of the graph $4x - 9y = -13$

26) What are the zeros of the function: $f(x) = x^3 + 7x^2 + 12x$?

27) The cost, in thousands of dollars, of producing x thousands of textbooks is C $(x) = x^2 + 5x + 15$. The revenue, also in thousands of dollars, is R $(x) = 6x$. Find the profit or loss if 3,000 textbooks are produced. (profit = revenue − cost)

28) Suppose a triangle has the dimensions indicated below; Then Sin B =?

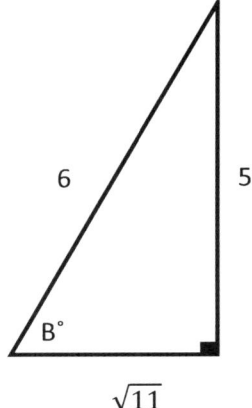

29) What is the solution of the following system of equations?

$$\begin{cases} 6x + y = 10 \\ 12x - 7y = -16 \end{cases}$$

30) Find the Center and Radius of the graph $(x - 2)^2 + (y + 4)^2 = $

18

31) Simplify $\dfrac{3 - 2i}{-3i}$

32) What is the domain of the following function?

$$f(x) = \sqrt{x - 3} + 5$$

33) Find the inverse function of $f(x) = \dfrac{x-2}{4}$

34) What is the value of x in the following equation?

$\log (x + 7) - \log (x - 2) = 1$

35) A number is chosen at random from 1 to 20. Find the probability of not selecting a composite number.

"The End of Test 1"

ALEKS Math Practice Test 2

Mathematics Placement Assessment

- ❖ **35 Questions.**

- ❖ **Total time for this test: No Limit Time.**

- ❖ **You may NOT use a calculator on this Section.**

Administered *Month Year*

1) Write the $\frac{3}{105}$ as a decimal.

2) $4 + 6 \times (-2) - [4 + 22 \times 5] \div 6 = ?$

3) If $6 + 2x \leq 16$, what is the value of $x \leq ?$

4) Simplify. $\dfrac{\dfrac{1}{3} - \dfrac{x+3}{6}}{\dfrac{x^2}{3} - \dfrac{1}{3}}$

5) A man owed $3,200 on his car. After making 49 payments of

$58 each, how much did he have left to pay?

6) Find the solutions of the following equation.

$$x^2 + 4x - 3 = 0$$

7) How many 3 × 3 squares can fit inside a rectangle with a height of 48 and width of 18?

8) $(x^4)^{\frac{5}{8}}$

9) What is 6352.48345 rounded to the nearest tenth?

10) Last Friday Jacob had $32.52. Over the weekend he received some money for cleaning the attic. He now has $55. How much money did he receive?

11) 25 is what percent of 50?

12) Liam's average (arithmetic mean) on two mathematics tests is 7. What should Liam's score be on the next test to have an overall of 8 for all the tests?

13) Find all values of x in this equation: $2x^2 + 7x + 3 = 0$

14) What is the value of x in this equation? $5^7 \times 5^6 = 7^x$

15) In the following triangle what is the value of x?

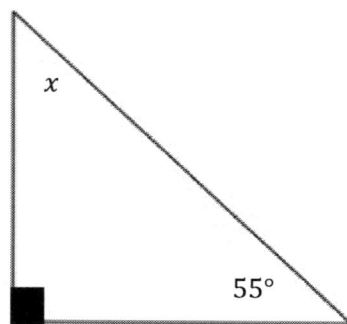

x

$55°$

16) Find the factors of $x^2 + 4x - 12$.

17) If a vehicle is driven 33 miles on Monday, 45 miles on Tuesday, and 30 miles on Wednesday, what is the average number of miles driven each day?

18) A ladder leans against a wall forming a 60° angle between the ground and the ladder. If the bottom of the ladder is 40 feet away from the wall, how long is the ladder?

19) What is the distance between the points $(2, -3)$ and $(5, 1)$?

20) $(x - 3)(x^2 + 4x + 3) =$?

21) What is the value of cos 45∘?

22) If θ is an acute angle and $\cos \theta = \frac{4}{5}$ then $\sin \theta$ =?

23) If $\log_2 x = 6$, then x = ?

24) If $f(x) = x - \frac{7}{3}$ and f^{-1} is the inverse of $f(x)$, what is the value of $f^{-1}(7)$?

25) Solve. $|10 - (16 \div |3 - 7|)|$ =?

26) What's the reciprocal of $\frac{x^3}{24}$?

27) What is the solution of the following system of equations?

$$\begin{cases} -2x - y = -9 \\ 5x - 3y = 9 \end{cases}$$

28) What is the equivalent temperature of 86°F in Celsius?

$C = \frac{5}{9} (F - 32)$

29) Simplify (−3 + 4i) (2 + 3i).

30) Find $\tan \frac{4\pi}{3}$

31) If $f(x) = 4x - 8$ and $g(x) = x^2 - 2x$, then find $(\frac{f}{g})(x)$.

32) What is the center and radius of a circle with the following equation?

$$(x - 4)^2 + (y + 7)^2 = 2$$

33) If the center of a circle is at the point $(-4, 3)$ and its circumference equals to 2π, what is the standard form equation of the circle?

34) Anita's trick–or–treat bag contains 20 pieces of chocolate, 16 suckers, 16 pieces of gum, 28 pieces of licorice. If she randomly pulls a piece of candy from her bag, what is the probability of her pulling out a piece of sucker?

35) What is the value of x in the following equation?

$$\log_4(x + 1) - \log_4(x - 1) = 1$$

"The End of Test 2"

Answers and Explanations
ALEKS Math Practice Tests
Answer Key

❋ Now, it's time to review your results to see where you went wrong and what areas you need to improve!

ALEKS Math Practice Test

Practice Test 1				Practice Test 2			
1	36	**21**	2,110	**1**	0.0286	**21**	$\sqrt{2}/2$
2	10	**22**	$-27 \leq x \leq 21$	**2**	-27	**22**	$3/5$
3	25	**23**	-1	**3**	$x \leq 5$	**23**	64
4	P^9	**24**	$3a^2\sqrt{b}$	**4**	$-1/2(x-1)$	**24**	$28/3$
5	$x \leq -5$	**25**	$4/9$	**5**	358	**25**	6
6	90	**26**	$0, -3, -4$	**6**	$-2 \pm \sqrt{7}$	**26**	$24/x^3$
7	$(x-3)$	**27**	21,000 loss	**7**	96	**27**	-5
8	250	**28**	$5/6$	**8**	$x^{\frac{5}{2}}$	**28**	30
9	32	**29**	1	**9**	6,352.5	**29**	$-(i+18)$
10	$x^2 + 8x + 15$			**10**	12.48	**30**	$\sqrt{3}$
11	4.8	**30**	$(2, -4), 3\sqrt{2}$	**11**	50	**31**	$4/x$
12	83°	**31**	$2/3 + i$	**12**	10	**32**	$(4, -7), \sqrt{2}$
13	4^3	**32**	$[3, +\infty)$	**13**	$-3, -1/2$	**33**	***
14	163.76	**33**	$2(2x+1)$	**14**	5^{13}	**34**	$1/5$
15	-9	**34**	3	**15**	35°	**35**	$5/3$
16	32	**35**	2	**16**	$(x-2)(x+6)$		
17	25%			**17**	36		
18	32			**18**	80		$***(x+4)^2 + (y-3)^2 = 1$
19	55			**19**	5		
20	$-x^2 - 4x - 6$			**20**	$x^3 + 2x^2 - 9x - 9$		

Answers and Explanations

ALEKS Mathematics Placement Assessment

Practice Tests 1

1) Answer: 36.

$\frac{28}{4} = 7, \frac{32}{4} = 8, \frac{36}{4} = 9$; 37,39 are prime and 38 not divisible by 4.

2) Answer: 10.

If $a = 5$ then $b = \frac{5^2}{5} + 5 \Rightarrow b = \frac{25}{5} + 5 \Rightarrow b = 5 + 5 = 10$

3) Answer: 25.

Let x be the number. Write the equation and solve for x.

$\frac{2}{3} \times 15 = \frac{2}{5} \cdot x \Rightarrow \frac{2 \times 15}{3} = \frac{2x}{5}$, use cross multiplication to solve for x.

$5 \times 30 = 2x \times 3 \Rightarrow 150 = 6x \Rightarrow x = 25$

4) Answer: P^9.

$(p^3) \cdot (p^6) = p^{3+6} = p^9$

5) Answer: $x \leq -5$.

$8 - \frac{2}{5}x \geq 10 \Rightarrow -\frac{2}{5}x \geq 2 \Rightarrow -x \geq 5 \Rightarrow x \leq -5$

6) Answer: 90.

$150 \times \frac{60}{100} = 90$

7) Answer: $(x - 3)$.

Factor each trinomial $x^2 - x - 6$ and $x^2 - 5x + 6$

$x^2 - x - 6 \Rightarrow (x - 3)(x + 2)$

$x^2 - 5x + 6 \Rightarrow (x - 2)(x - 3) \Rightarrow$ The common Factor is: $(x - 3)$

8) Answer: 250.

Th ratio of boy to girls is 5:3. Therefore, there are 5 boys out of 8 students. To find the answer, first divide the total number of students by 8, then multiply the result by 5.

$400 \div 8 = 50 \Rightarrow 50 \times 5 = 250$

9) Answer: 32.

$a = 8 \Rightarrow$ area of triangle is $= \frac{1}{2}(8 \times 8) = \frac{64}{2} = 32$ cm

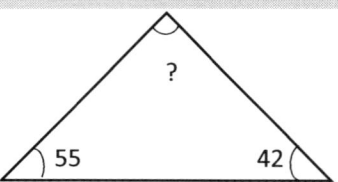

Isosceles right triangle

10) Answer: $x^2 + 8x + 15$.

Use FOIL (First, Out, In, Last)

$(x + 3)(x + 5) = x^2 + 5x + 3x + 15 = x^2 + 8x + 15$

11) Answer: 4.8.

$\frac{6+5b}{30b^2} = \frac{1}{b^2} \Rightarrow (b \neq 0)\ b^2(6 + 5b) = 30b^2 \Rightarrow 6 + 5b = 30 \Rightarrow 5b = 24 \Rightarrow b = 4.8$

12) Answer: 83°

$55° + 42° = 97°$

$180° - 97° = 83°$

The value of the third angle is 83°.

13) Answer: 4^3.

$4^{\frac{7}{3}} \times 4^{\frac{2}{3}} = 4^{\frac{7}{3}+\frac{2}{3}} = 4^{\frac{9}{3}} = 4^3$

14) Answer: 163.76.

Underline the hundredths place: 163.7<u>5</u>99

Look to the right if it is 5 or bigger, add it by 1.

Then, round up to 163.76

15) Answer: −9.

$x^2 - 81 = 0 \Rightarrow x^2 = 81 \Rightarrow x = 9$ or $x = -9$

16) Answer: 32.

If $a = 2$, then:

$3a^2 + 4a + 12 \Rightarrow 3(2)^2 + 4(2) + 12 \Rightarrow 3(4) + 8 + 12 = 32$

17) Answer: 25%.

The question is this: 469.50 is what percent of 626?

Use percent formula: $\text{part} = \dfrac{\text{percent}}{100} \times \text{whole}$

$469.50 = \dfrac{\text{percent}}{100} \times 626 \Rightarrow 469.50 = \dfrac{\text{percent} \times 626}{100}$

$\Rightarrow 46950 = \text{percent} \times 626 \Rightarrow \text{percent} = \dfrac{46950}{626} = 75$

530.40 is 85 % of 624. Therefore, the discount is: $100\% - 75\% = 25\%$

18) Answer: 32.

Let x be the smallest number. Then, these are the numbers:

$x, x + 1, x + 2, x + 3, x + 4$

$\text{average} = \dfrac{\text{sum of terms}}{\text{number of terms}} \Rightarrow 34 = \dfrac{x+(x+1)+(x+2)+(x+3)+(x+4)}{5} \Rightarrow 34 =$

$\dfrac{5x+10}{5} \Rightarrow 170 = 5x + 10 \Rightarrow 160 = 5x \Rightarrow x = 32$

19) Answer: 55.

$a + b + c = 65$

$\dfrac{a+b+c+d}{4} = 30 \Rightarrow a + b + c + d = 120 \Rightarrow 45 + d = 120$

$d = 120 - 65 = 55$

20) Answer: $-x^2 - 4x - 6$.

$(g - f)(x) = g(x) - f(x) = (-x^2 - 2 - 3x) - (4 + x)$

$-x^2 - 2 - 3x - 4 - x = -x^2 - 4x - 6$

21) Answer: 2,110.

$42 \times \$125 = \$5250$

Payable amount is: $\$7360 - \$5250 = \$2110$

22) Answer: $-27 \le x \le 21$.

$\frac{|3+x|}{8} \le 3 \Rightarrow |3 + x| \le 24 \Rightarrow -24 \le 3 + x \le 24 \Rightarrow -24 - 3 \le$

$x \le 24 - 3 \Rightarrow -27 \le x \le 21$

23) Answer: -1.

$\tan\left(-\frac{\pi}{4}\right) = -1$

24) Answer: $3a^2\sqrt{b}$.

$\frac{\sqrt{27a^5b^3}}{\sqrt{3ab^2}} = \frac{3a^2b\sqrt{3ab}}{b\sqrt{3a}} = 3a^2\sqrt{b}$

25) Answer: $\frac{4}{9}$.

$-9y = -4x - 13 \Rightarrow y = \frac{-4}{-9}x - \frac{13}{-9} \Rightarrow y = \frac{4}{9}x + \frac{13}{9}$

26) Answer: $0, -3, -4$.

Frist factor the function:

$x(x + 3)(x + 4)$

To find the zeros, $f(x)$ should be zero.

$f(x) = x(x + 3)(x + 4) = 0$

Therefore, the zeros are: $x = 0$

$(x + 3) = 0 \Rightarrow x = -3$

$(x + 4) = 0 \Rightarrow x = -4$

27) Answer: 21,000 loss.

$c(3) = (3)^2 + 5(3) + 15 = 9 + 15 + 15 = 39$

$6 \times 3 = 18 \Rightarrow 18 - 39 = -21 \Rightarrow 21,000 \text{ loss}$

28) Answer: $\frac{5}{6}$.

$$\sin B = \frac{\text{Opposite Side}}{\text{Hypotenuse}} = \frac{5}{6}$$

29) Answer: 1.

$\begin{cases} 6x + y = 10 \\ 12x - 7y = -16 \end{cases}$ $\Rightarrow$ Multiply the first equation by -2 $\Rightarrow$

$\begin{cases} -12x - 2y = -20 \\ 12x - 7y = -16 \end{cases}$

Add two equations $\Rightarrow -9y = -36 \Rightarrow y = 4$ then: $x = 1$

30) Answer: $(2, -4), 3\sqrt{2}$.

$(x - h)^2 + (y - k)^2 = r^2$ $\Rightarrow$ center: (h, k) and radius: r

$(x - 2)^2 + (y + 4)^2 = 18 \Rightarrow$ center: $(2, -4)$ and radius: $3\sqrt{2}$

31) Answer: $\frac{2}{3} + i$.

$$\frac{3 - 2i}{-3i} \times \frac{i}{i} = \frac{3i - 2i^2}{-3i^2} = \frac{3i - 2(-1)}{-3(-1)} = \frac{2}{3} + i$$

32) Answer: $[3, +\infty)$

The number under the square root symbol must be zero or greater

than zero therefore: $x - 3 \geq 0 \Rightarrow x \geq 3$

domain of function$= [3, +\infty)$

33) Answer: $2(2x + 1)$.

$f(x) = \frac{x-2}{4} \Rightarrow y = \frac{x-2}{4} \Rightarrow 4y = x - 2 \Rightarrow 4y + 2 = x$

$f^{-1} = 4x + 2 = 2(2x + 1)$

34) Answer: 3.

<u>METHOD ONE:</u>

$\log(x + 7) - \log(x - 2) = 1$, Add $\log(x - 2)$ to both sides:

$\log (x + 7) - \log (x - 2) + \log (x - 2) = 1 + \log (x - 2)$

And simplify: $\log (x + 7) = 1 + \log (x - 1)$

Logarithm rule: $a = \log_b (b^a) \Rightarrow 1 = \log_{10}(10^1) = \log (10)$

then: $\log (x + 7) = \log (10) + \log_2 (x - 2)$

Logarithm rule: $\log_c (a) + \log_c (b) = \log_c (ab)$

then: $\log (10) + \log (x - 2) = \log (10(x - 2))$

$\log (x + 7) = \log (10(x - 2))$

When the logs have the same base:

$\log_b (f(x)) = \log_b (g(x)) \Rightarrow f(x) = g(x)$

$x + 7 = 10(x - 2) \Rightarrow x + 7 = 10x - 20 \Rightarrow 9x = 27 \Rightarrow x = 3$

METHOD TWO

We know that: $\log_a b - \log_a c = \log_a \frac{b}{c}$, and $\log b = c \Rightarrow b = 10^c$

Then, $\log (x + 7) - \log (x - 2) = \log \frac{x+7}{x-2} = 1 \Rightarrow \frac{x+7}{x-2} = 10^1 \Rightarrow x +$

$7 = 10(x - 2) \Rightarrow x + 7 = 10x - 20 \Rightarrow 10x - x = 7 + 20 \Rightarrow x = 3$

35) Answer: 2.

Set of number that are not composite between 1 and 20: A= {1, 2, 5, 7, 11, 13, 17, 19}

$n(A) = 10 \Rightarrow p = \dfrac{10}{20} = 2$

Answers and Explanations

ALEKS Mathematics Placement Assessment

Practice Tests 2

1) Answer: 0.0286.

$$\frac{3}{105} = \frac{1}{35} = 0.02857143 \cong 0.0286$$

2) Answer: -27.

Use PEMDAS (order of operation):

$4 + 6 \times (-2) - [4 + 22 \times 5] \div 6 = 4 + (-12) - [4 + 110] \div 6 = -8 - 114 \div 6$

$= -8 - 19 = -27.$

3) Answer: $x \leq 5$.

$6 + 2x \leq 16 \Rightarrow 2x \leq 16 - 6 \Rightarrow 2x \leq 10 \Rightarrow x \leq 5$

4) Answer: $-\dfrac{1}{2(x-1)}$.

Simplify:

$$\frac{\frac{1}{3} - \frac{x+3}{6}}{\frac{x^2}{3} - \frac{1}{3}} = \frac{\frac{1}{3} - \frac{x+3}{6}}{\frac{x^2-1}{3}} \quad \left(\text{Simplify: } \frac{1}{3} - \frac{x+3}{6} = \frac{-x-1}{6}\right)$$

then: $\dfrac{\frac{-x-1}{6}}{\frac{x^2-1}{3}} = \dfrac{-x-1}{6} \times \dfrac{3}{x^2-1} = \dfrac{-3(x+1)}{6(x-1)(x+1)} = -\dfrac{1}{2(x-1)}$

5) Answer: $358.

$49 \times \$58 = \$2842$ Payable amount is: $\$3200 - \$2842 = \$358$

6) Answer: $-2 \pm \sqrt{7}$.

$$X_{1,2} = \frac{-b \pm \sqrt{b^2 - 4ac}}{2a}$$

$ax^2 + bx + c = 0$

$x^2 + 4x - 3 = 0 \Rightarrow$ then: a = 1, b = 4 and c = − 3

$x = \dfrac{-4 + \sqrt{4^2 - 4.1.-3}}{2.1} = -2 + \sqrt{7}$; $x = \dfrac{-4 - \sqrt{4^2 - 4.1.-3}}{2.1} = -2 - \sqrt{7}$

7) Answer: 96.

Number of squares equal to: $\dfrac{48 \times 18}{3 \times 3} = 16 \times 6 = 96$

8) Answer: $x^{\frac{5}{2}}$.

$(x^4)^{\frac{5}{8}} = x^{4 \times \frac{5}{8}} = x^{\frac{20}{8}} = x^{\frac{5}{2}}$

9) Answer: 6,352.5.

Underline the tenth place:6352.48345

Look to the right if it is 5 or bigger, add 1 to the underlined digit.

Then, round up the decimal to 6352.5

10) Answer: $12.48.

$55 - $32.52 = $12.48

11) Answer: 50.

$50 \times \dfrac{x}{100} = 25 \Rightarrow 50 \times x = 2500 \Rightarrow x = \dfrac{2500}{50} = 50$

12) Answer: 10.

$\dfrac{a+b}{2} = 7 \Rightarrow a + b = 14$

$\dfrac{a+b+c}{3} = 8 \Rightarrow a + b + c = 24$

$14 + c = 24 \Rightarrow c = 24 - 14 = 10$

13) Answer: −3, −1/2.

$x_{1,2} = \dfrac{-b \pm \sqrt{b^2 - 4ac}}{2a}$

$ax^2 + bx + c = 0$

$2x^2 + 7x + 3 = 0$ ⇒then: a = 2, b = 7 and c = 3

$$X = \frac{-7 + \sqrt{7^2 - 4.2.3}}{2.2} = -\frac{1}{2}$$

$$X = \frac{-7 - \sqrt{7^2 - 4.2.3}}{2.2} = -3$$

14) Answer: 5^{13}.

$5^7 \times 5^6 = 5^{7+6} = 5^{13}$

15) Answer: $35°$.

90° + 55° = 145°

180° - 145° = 35°

16) Answer: $(x - 2)(x + 6)$.

$x^2 + 4x - 12 = (x - 2)(x + 6)$

17) Answer: 36.

33 + 45 + 30 = 108

Average = $\frac{108}{3}$ = 36

18) Answer: 80.

The relationship among all sides of special right triangle

$30° - 60° - 90°$ is provided in this triangle:

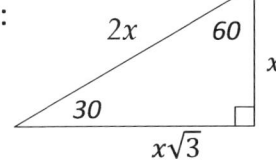

In this triangle, the opposite side of 30° angle is half of the

hypotenuse.

Draw the shape of this question:

The latter is the hypotenuse. Therefore, the latter is 80 ft.

19) Answer: 5.

$$C = \sqrt{(x_A - x_B)^2 + (y - y_B)^2}$$

$$C = \sqrt{(2 - (5))^2 + (-3 - 1)^2}$$

$$C = \sqrt{(3)^2 + (-4)^2} \Rightarrow C = \sqrt{9 + 16}$$

$$\Rightarrow C = \sqrt{25} = 5$$

20) Answer: $x^3 + x^2 - 9x - 9$.

Use FOIL (First, Out, In, Last)

$$(x - 3)(x^2 + 4x + 3) = x^3 + 4x^2 + 3x - 3x^2 - 12x - 9$$

$$= x^3 + x^2 - 9x - 9$$

21) Answer: $\dfrac{\sqrt{2}}{2}$.

$$cos\ 45° = \frac{\sqrt{2}}{2}$$

22) Answer: $\dfrac{3}{5}$.

$cos\theta = \dfrac{4}{5} \Rightarrow$ we have following triangle, then

$$c = \sqrt{5^2 - 4^2} = \sqrt{25 - 16} = \sqrt{9} = 3$$

$$sin\theta = \frac{3}{5}$$

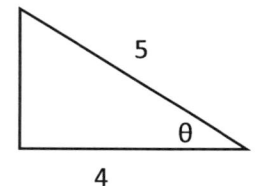

23) Answer: 64.

METHOD ONE:

$\log_2 x = 6$

Apply logarithm rule: $a = \log_b(b^a)$

$6 = \log_2(2^6) = \log_2(64)$

$\log_2 x = \log_2(64)$

When the logs have the same base:

$$\log_b(f(x)) = \log_b(g(x)) \Rightarrow f(x) = g(x)$$

then: $x = 64$

METHOD TWO:

We know that:$\log_a b = c \Rightarrow b = a^c \Longrightarrow \log_2 x = 6 \Rightarrow x = 2^6 = 64$

24) **Answer:** $\frac{28}{3}$.

$$f(x) = x - \frac{7}{3} \Rightarrow y = x - \frac{7}{3} \Rightarrow y + \frac{7}{3} = x$$

$$f^{-1}(x) = x + \frac{7}{3} \Longrightarrow f^{-1}(7) = 7 + \frac{7}{3} = \frac{28}{3}$$

25) **Answer: 6.**

$$|10 - (16 \div |3 - 7|)| = |10 - (16 \div |-4|)| = |10 - (16 \div 4)| =$$
$$|10 - 4| = |6| = 6$$

26) **Answer:** $\frac{24}{x^3}$.

$$\frac{x^3}{24} \Rightarrow \text{reciprocal is: } \frac{24}{x^3}$$

27) **Answer:** -5.

$$\begin{cases} -2x - y = 9 \\ 5x - 3y = 5 \end{cases} \Rightarrow \text{Multiplication} - 3 \text{ in first equation}$$

$$\Rightarrow \begin{cases} 6x + 3y = -27 \\ 5x - 3y = 5 \end{cases}$$

Add two equations together $\Rightarrow 11x = -22 \Rightarrow x = -2$

then: $y = -5$

28) **Answer: 30.**

Plug in 86 for F and then solve for C.

$$C = \frac{5}{9}(F - 32) \Rightarrow C = \frac{5}{9}(86 - 32) \Rightarrow C = \frac{5}{9}(54) = 30$$

29) **Answer:** $-(i + 18)$.

We know that: $i = \sqrt{-1} \Rightarrow i^2 = -1$

$$(-3 + 4i)(2 + 3i) = -6 - 9i + 8i + 12i^2 = -6 - i - 12 = -i - 18$$

30) Answer: $\sqrt{3}$.

$$\tan\frac{4\pi}{3} = \frac{\sin\frac{4\pi}{3}}{\cos\frac{4\pi}{3}} = \frac{\frac{\sqrt{3}}{2}}{\frac{1}{2}} = \sqrt{3}$$

31) Answer: $\frac{4}{x}$.

$$\left(\frac{f}{g}\right)(x) = \frac{f(x)}{g(x)} = \frac{4x - 8}{x^2 - 2x} = \frac{4(x-2)}{x(x-2)} = \frac{4}{x}$$

32) Answer: $(4, -7)$, $\sqrt{2}$.

$$(x - h)^2 + (y - k)^2 = r^2 \ \Rightarrow \text{center: } (h, k) \text{ and radius: } r$$

$$(x - 4)^2 + (y + 7)^2 = 2 \ \ \Rightarrow \text{center: } (4, -7) \text{ and radius: } \sqrt{2}$$

33) Answer: $(x + 4)^2 + (y - 3)^2 = 1$.

Use formula of a circle in the coordinate plane:

$$(x - h)^2 + (y - k)^2 = r^2 \ \Rightarrow \text{center: } (h, k) \text{ and radius: } r$$

center: $(-4, 3)$ $\Rightarrow h = -4, k = 3$

circumference $= 2\pi \Rightarrow$ circumference $= 2\pi r = 2\pi \Rightarrow r = 1$

$$(x + 4)^2 + (y - 3)^2 = 1$$

34) Answer: $\frac{1}{5}$.

$$\text{Probability} = \frac{number\ of\ desired\ outcomes}{number\ of\ total\ outcomes} = \frac{16}{20+16+16+28} = \frac{16}{80} = \frac{1}{5}$$

35) Answer: $\frac{5}{3}$.

METHOD ONE

$$\log_4(x + 1) - \log_4(x - 1) = 1$$

Add $\log_4(x - 1)$ to both sides

$$\log_4(x + 1) - \log_4(x - 1) + \log_4(x - 1) = 1 + \log_4(x - 1)$$

$$\log_4(x + 1) = 1 + \log_4(x - 1)$$

Apply logarithm rule: $a = \log_b(b^a) \Rightarrow 1 = \log_4(4^1) = \log_4(4)$

then: $\log_4(x + 1) = \log_4(4) + \log_4(x - 1)$

Logarithm rule: $\log_c(a) + \log_c(b) = \log_c(ab)$

then: $\log_4(4) + \log_4(x - 1) = \log_4(4(x - 1))$

$\log_4(x + 1) = \log_4(4(x - 1))$

When the logs have the same base:

$\log_b(f(x)) = \log_b(g(x)) \Rightarrow f(x) = g(x)$

$(x + 1) = 4(x - 1) \Rightarrow x = \dfrac{5}{3}$

METHOD TWO:

We know that: $\log_a b - \log_a c = \log_a \dfrac{b}{c}$ and $\log_a b = c \Rightarrow$

$b = a^c$

Then: $\log_4(x + 1) - \log_4(x - 1) = \log_4 \dfrac{x+1}{x-1} = 1 \Rightarrow \dfrac{x+1}{x-1} =$

$4^1 = 4 \Rightarrow x + 1 = 4(x - 1) \Rightarrow x + 1 = 4x - 4 \Rightarrow 4x - x = 1 +$

$4 \rightarrow 3x = 5 \Rightarrow x = \dfrac{5}{3}$

"End"

Made in the USA
San Bernardino,
CA